D0765545

worldwide selection of EXOTIC fruits and vegetables

worldwide collection of

EXOTIC fruits and vegetables

Julia Richardson

Données de catalogage avant publication (Canada)

Richardson, Julia

The worldwide selection of exotic fruits and vegetables

ISBN 2-7625-6052-7

1. Cookery (Fruit). 2. Cookery (Vegetables). 3. Fruit. 4. Vegetables. I. Title.

TX811.R52 1990 641.3'4 C90-096621-1

PRODUCTION
Ginette Guétat
Author:
Julia Richardson
Co-author:
Michel Brassard
Consultants:
Fruits Botner Ltée
Holly Botner
Louise Hébert
Nutritionist:
Michel Pépin M.D., M. (Nut.)
Julie Bélanger, dt. p.

ART AND DESIGN
Christiane Litalien
Desktop publishing:
Marc Chapdelaine

PHOTOGRAPHY
Paul Casavant
Food stylists:
Marie-Lise Dubuc
Lucie Casavant
Denise Dion

PUBLISHED BY
Les Éditions Héritage Inc.
300, rue Arran
Saint-Lambert, Québec J4R 1K5
Tél: (514) 875-0327

President:
Jacques Payette

Marketing
Luc Payette
Assistant: Johanne Phaneuf

Dishes graciously provided by
Le Cache-Pot, 5047, St-Denis St., Montréal

© 1990 LES ÉDITIONS HÉRITAGE INC.

All rights reserved

Legal deposit: 4th quarter 1990
National Library of Canada

ISBN : 2-7625-6052-7 Printed in Canada

Table of Contents

Foreword

This book is dedicated to all the fruit and vegetable growers and distributors throughout the world, and to adventuresome people everywhere who love to try new things. A greater variety of fresh produce than ever before is becoming widely available in supermarkets and fruit and vegetable stores. I hope this book provides a useful guide for those who are looking at an item for the first time, wondering what on earth to do with it.

Do you ever look at a mangosteen and think to yourself that perhaps a few days ago it was ripening on a tree in far-off Thailand? Have you ever imagined the succulent papaya you are eating growing in the exotic tropics of Hawaii or Brazil? The discovery of new fruits and vegetables is of great interest to me as a lover of good food and a keen traveller. I've been lucky enough to have visited about forty countries on five continents and tasting exotic new foods has always been a very enjoyable part of the experience. Fruits and vegetables are second nature to me, as a member of the fourth generation of my family in the produce business. Having grown up in a household where salads are eaten for breakfast, lunch and supper and where no one thinks it unusual to see half a dozen vegetables in one meal or over a dozen varieties of fruit offered for dessert, I definitely feel at home with produce from around the world.

The fresh produce market is constantly changing. As a fruit or vegetable becomes more popular, cultivators in different countries start to grow it and, therefore their availability changes and new varieties are developed. Suddenly, it seems that a fruit you knew from Asia that is egg-sized and green and available only for two months a year, is now grown in a dozen other countries and comes in three colours, and six sizes and can be found for eight months a year.

My grateful thanks to Eric Botner, the guiding force behind this book, to Holly Botner who worked on this project from the beginning, to Louise Hébert, the Director of Marketing and Promotion at Fruits Botner Ltée for her invaluable expertise, to all the Fruits Botner personnel who contributed to the success of this project, to Christiane Litalien and Marc Chapdelaine for their excellent work on the layout, to Paul Casavant for his photographs and to Ginette Guétat and Jacques Payette of Les Éditions Héritage Inc.

Most of the material in this book was originally published as a series of seven magazines. If you have difficulty locating specific items (because we have followed the same order of presentation in the book as we did in the magazines), you will find the index very helpful.

Enjoy experimenting!

Julia Richardson

Introduction to Oriental Vegetables

Oriental cuisine — including Chinese, Japanese, Vietnamese, Burmese, Korean, Laotian, Thai, Indonesian and Malaysian cookery — is a vast subject. We can only hope to introduce our readers to a variety of the oriental vegetables that play important roles within this broad category. Some of these exotic vegetables are becoming more widely available at supermarkets and fresh produce stores.

Many oriental recipes call for stir-frying. If possible, use a wok or a deep cast-iron skillet for stir-fries in order to cook the food quickly and evenly, and heat the wok before adding oil. The best oil for stir-frying is peanut oil, though you can use soybean or corn oil. Oriental sesame seed oil enhances most dishes with its delicious, nutty flavour, but it is not very good for cooking, as it burns easily; a few drops are added to a dish just before serving.

A bottle of rice wine is useful to have on hand; one of the best is called Shao Hsing, Hsao Shing or Shaoxing. While you can substitute sherry, the flavour is not the same. Try using Chinese five-spice powder — a mixture made from cloves, star anise, cinnamon, fennel seeds and Szechwan pepper corns — in barbecue sauces and marinades. You may also wish to try some Chinese vinegars; the white one is usually used in sweet and sour dishes, the red one in stir-fries and the black one in braising, or as a table condiment. Another choice to make is between dark and light soy sauce; the dark one is sweeter, the light saltier.

A number of bottled sauces can add marvelous flavours to your cooking. The popular hoisin sauce is thick, dark brown, sweet and spicy. Fermented bean curd black bean paste, yellow bean paste, oyster sauce and fish sauce all add wonderful flavour to stir-fried food. You might also like to try miso — a Japanese fermented bean paste — peanut butter or sesame butter in your sauces. Toasted sesame seeds, raw peanuts, pine nuts, cashews and almonds make excellent garnishes for many oriental dishes. Fresh bean curd (tofu), rather bland in taste, is a good source of protein in vegetable dishes.

The special ingredients necessary to oriental cooking are easy to come by at Chinese, Japanese and other ethnic or specialty grocery stores. Many oriental vegetables can be used in western recipes, though others are best complemented by a variety of oriental spices and sauces.

Bean sprouts, which have become commonplace in many supermarkets, combine well with nearly every oriental vegetable. Some of the less commonly available fresh vegetables to look for are baby corn, bamboo shoots, Chinese spinach (amaranth), water spinach, mustard greens and chrysanthemum leaves. Scallions, shallots, leeks and garlic are frequently used and lightly cooked watercress adds a good tang to many dishes.

Enjoy experimenting!

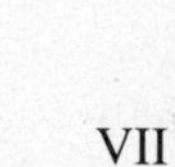

Pineapple

LOW IN CALORIES
EXCELLENT SOURCE OF MINERALS
GOOD SOURCE OF VITAMIN C
EXCELLENT FOR DIGESTION

Origin

• The pineapple, the most commercially valuable plant in the Bromeliaceæ family, originated in the region where Brazil, Argentina and Paraguay adjoin. The Spanish called this luscious fruit "pina" — or pine cone, because of its appearance — hence the word "pineapple". The Guarani Indian name, "nana ment", meaning "exquisite fruit", gave rise to the French word "anana".

• Pineapples were well known in most of Latin America long before the arrival of Columbus. Much appreciated by explorers to the New World, they were introduced to Portuguese and Spanish colonies in Asia before the end of the 16th century. Improved varieties were developed in Europe for cultivation in the greenhouses of the rich. By the 19th century, commercial cultivation had begun in the Portuguese Azores, off the northwest African coast, as well as in Hawaii, but not until modern transportation and refrigeration were developed did the pineapple come into its own. Pineapples require rapid shipment to market, preferably by air, as they are ripe when harvested and therefore fragile.

• Today, Hawaii is a major producer, along with Mexico, Central and South America, the Caribbean islands and a number of African and Asian countries.

Varieties and description

• Many pineapple hybrids exist, deriving from four major varieties: the Smooth Cayenne, the Queen, the Red Spanish and the Abacaxi. They differ in size of fruit and crown, shape, skin and flesh colour, and have varying levels of sweetness and acidity. The rough pineapple skin can be green, golden, reddish-orange or brown and its flesh ranges from pale yellow to a deep golden colour. Some varieties have a wonderful tropical aroma.

• A pineapple is made of dozens of true fruits, each of which develops from one of the individual flowers that merge around the central stem, or core. Its spikey crown of rigid leaves grows from the top of the core.

• One excellent new variety from Taiwan and Southeast Asia, called the cherimoya pineapple, is only rarely seen in North America. This sweet and juicy elongated pineapple is easily pulled apart by hand in small sections.

Availability

• Pineapples can be found year round, from various producer countries.

Selection

• Pineapples are perishable and need careful handling. Choose one that feels heavy for its size and has a fresh-looking crown. Avoid those with soft or moist patches.

Storage

• Pineapples are picked ripe and should be eaten as soon as possible; they deteriorate if left sitting around. Keep at room temperature for one or two days, or refrigerate in a plastic bag and use within a few days.

• Pineapples keep longest when the fruit is cut up, placed in an airtight container with as much of its juice as possible, and refrigerated. It can be frozen this way, though it tastes best when eaten fresh.

Preparation

• Pineapples can be cut in a variety of decorative ways. The easiest method is to cut off the crown and base, stand it upright and slice off the skin in vertical strips. Use a sharp, pointed knife to pick out any remaining "eyes". The pineapple can then be cut into rings or into quarters lengthwise, which makes it simple to cut away the hard core. Some supermarkets have coring machines for customer convenience.

• Pineapples contain the enzyme bromeline, which breaks down protein and therefore is a good meat tenderizer and a digestive aid. Bromeline prevents gelatin from setting, unless the pineapple is cooked first. It also curdles milk products if combined with them and left standing too long before serving.

Suggestions

• The versatile pineapple fits in well at any stage of a meal or alone, as a snack. Try fresh pineapple juice or cocktails, pineapple appetizers, pineapple in salads, pineapple main courses — especially with seafood, chicken, pork and stir-fried dishes — and pineapple desserts.

• Grill pineapple rings for a few minutes each side and serve with roasted meat or poultry.

• Try cooking rice in half stock and half pineapple juice. Heat a little oil in a deep frying pan and sauté finely chopped onions, celery and slivered peppers, adding some finely chopped pineapple, cooked peas and the cooked rice after a minute or two. Stir for a few minutes, seasoning to taste and serve with ribs, roast chicken, kebabs, or the meal of your choice.

• Chopped pineapple is great in cottage cheese, cole slaw, chef's salad with chunks of ham and cheese, shrimp salad, chicken salad or fruit salad.

• For a great sweet and sour sauce for chicken: Combine 250 mL (1 cup) pineapple chunks with 60 mL (1/4 cup) lime juice, 30 mL (2 TBSP) each brown sugar, molasses, soy sauce and tomato paste, 15 mL (1 TBSP) each fresh grated ginger and cornstarch, 5 mL (1 tsp) dry mustard, 2 minced garlic cloves, 1 finely chopped onion and a little salt and pepper. Arrange chicken legs and wings in a greased baking dish and pour the sauce over them. Let marinate for an hour, then bake in a moderately hot oven for 40 to 50 minutes, turning the chicken once or twice.

• Pineapple is excellent in drinks and desserts prepared with rum.

• Grow an attractive houseplant from your pineapple crown: Clean off any remaining flesh from the base of the crown, exposing the root buds. Break off a few bottom leaves and set the crown aside for a week in a dry, shaded place. Plant in soil in a pot with good drainage, water once a week; fertilize at planting and every 2 months afterward. Transplant to a bigger pot when necessary.

Nutritional Value

Pineapple

(per 155 g = 5 1/2 oz = 1 cup)

			RDA**	%
calories*	(kcal)	77		
protein*	(g)	0.6		
carbohydrates*	(g)	19.2		
fat*	(g)	0.7		
fibre	(g)	0.8	30	2.7
vitamin A	(IU)	35	3300	1.1
vitamin C	(mg)	24	60	40
thiamin	(mg)	0.14	0.8	17.5
riboflavin	(mg)	0.06	1.0	6
niacin	(mg)	0.7	14.4	4.9
sodium	(mg)	1	2500	0.04
potassium	(mg)	175	5000	3.5
calcium	(mg)	11	900	1.2
phosphorus	(mg)	11	900	1.2
magnesium	(mg)	21	250	8.4
iron	(mg)	0.57	14	4.1
water	(g)	135.1		

*RDA variable according to individual needs
**RDA recommended dietary allowance (average daily amount for a normal adult)

Golden flesh ↑

↓ Cored pineapple

Pineapple surprise

Ingredients		
1	fresh pineapple	
125 mL	strawberries, quartered	1/2 cup
125 mL	white rum	1/2 cup
500 mL	vanilla ice cream	2 cups
500 mL	whipped 35 % cream, slightly sweetened	2 cups

Instructions

- Cut the pineapple in half lengthwise, including the tuft of leaves.
- Carefully cut the flesh from each half, using a very sharp knife, leaving the shell intact.
- Chop the flesh in chunks, mix together with the strawberries, and add the rum. Refrigerate for at least an hour.
- To serve, fill the hollowed shells with alternating layers of mixed fruit and vanilla ice cream.
- Cover with whipped cream and serve on a bed of crushed ice.

Mangosteen

LOW IN CALORIES
EXCELLENT SOURCE OF FIBRE
CONTAINS SOME VITAMINS AND MINERALS

Origin

• The mangosteen, originally from Malaysia, belongs to the Guttiferæ family, along with several little-known tropical fruits, including the mundu and the San Domingo apricot. This delectable, exotic fruit is quite difficult to grow. The soil and climate must be perfectly suitable for cultivation; not only does it take from 10 to 15 years for a tree to produce fruit, but each tree has a good yield only every second year.

• Easy to digest and soothingly refreshing, the mangosteen was once prescribed by colonial doctors in Southeast Asia for fevers, dysentery and other ailments.

• Thailand, Indonesia and the Philippines are the biggest producers of mangosteens, though some are grown in India and Sri Lanka. Growers in Brazil, the Caribbean and Australia are attempting to introduce them as a commercial crop.

Varieties and description

• Although there are more than 100 varieties of mangosteens — many grown for ornamental purposes as they produce beautiful flowers — only one variety is grown for export.

• About 6 centimetres (2 1/2 inches) in diametre, the mangosteen is round and has a smooth, leathery, dark purple skin and a short woody stem surrounded by four rigid leaves (sepals). Under the skin, a very thick, pinkish-purple inedible membrane surrounds its white, juicy, sweet and fragrant flesh. The luscious flesh is divided into four to eight segments — two or three of which contain seeds that are edible. Though the taste may be compared to lychees or rambutans (both of which are not related to the mangosteen), the flavour is superior.

Availability

• Mangosteens from Southeast Asia are in season from May to September. Brazil is beginning to grow mangosteens for export in the winter months.

Selection

• Mangosteens are harvested only when perfectly ripe and ready to eat. They must be flown to market, as they are highly perishable. Mangosteens should be firm, never hard, and yield slightly to pressure.

Storage

• Mangosteens are best when eaten as fresh as possible. They can be left at room temperature for two to three days, or refrigerated for up to a week.

Preparation

• These exceptional fruits are finest eaten just as they are. Simply make a slit about 1 centimetre (1/3 inch) deep around the middle of each fruit and twist off the top. Use fingers or a fork to pick out the mangosteen segments, being careful not to stain your clothing with the pith.

• Or break open the skin with your thumbs by pressing lightly, then peel away the outer shell.

Suggestions

• Mangosteens served in the half shell make an attractive and delicious garnish for a variety of dishes and desserts.

• Decorate cakes, ice cream or fruit salad with mangosteen segments.

Praline-mangosteen cake icing

Ingredients		
375 mL	icing sugar	$1^1/_2$ cups
75 mL	butter, softened	1/3 cup
30 mL	praline*	2 TBSP
	pulp of 12 mangosteens, puréed	

Instructions

- Sift the icing sugar into a large bowl, add butter and beat until light and fluffy.
- Stir in praline, then gradually add the mangosteen pulp.
- Refrigerate one hour, then use to ice a cake.

** Praline is made with almonds or hazelnuts and is added to desserts for flavouring. Buy it ready-made, or try the following: Toast 250 grams (1/2 lb) slivered almonds. Over high heat, combine 250 mL (1 cup) fine sugar with 15 mL (1 TBSP) water. When the sugar begins to bubble, add a few drops of vanilla and the almonds, then stir briskly for a minute and a half. Remove from heat, spread mixture on an oiled baking sheet and cool for an hour. Break into chunks and crush until fine.*

Royal mangosteen cocktail

Ingredients		
6-7	mangosteens	
60 mL	sugar	1/4 cup
	raspberry liqueur (Framboise)	
1	bottle of champagne	

Instructions

- Break mangosteens into segments; sprinkle them with sugar and refrigerate for 1 hour.
- For each cocktail, put a few segments of fruit and syrup into each glass, a dash of Framboise and top with chilled champagne.

Nutritional Value
Mangosteen
(per 100 g = $3^1/_2$ oz)

			RDA**	%
calories*	(kcal)	57		
protein*	(g)	0.5		
carbohydrates*	(g)	14.7		
fat*	(g)	0.3		
fibre	(g)	5	30	16.7
vitamin A	(IU)	0	3300	0
vitamin C	(mg)	4	60	6.7
thiamin	(mg)	0.03	0.8	3.8
riboflavin	(mg)	0.02	1.0	2
niacin	(mg)	0.6	14.4	4.2
sodium	(mg)	1	2500	0.04
potassium	(mg)	136	5000	2.7
calcium	(mg)	10	900	1.1
phosphorus	(mg)	10	900	1.1
magnesium	(mg)		250	
iron	(mg)	0.5	14	3.6
water	(g)	84.3		

*RDA variable according to individual needs
**RDA recommended dietary allowance (average daily amount for a normal adult)

Kiwi

EXCELLENT SOURCE OF VITAMIN C
LOW IN CALORIES
GOOD SOURCE OF MINERALS

Origin

• Kiwi fruit, the name New Zealanders dreamed up for what was once known as the Chinese gooseberry, originated in southern China, where it was called yangtao. It is the most important of the three edible fruits in the Actinidiæ family; the others are not known outside Japan and Russia. In China, kiwis have been cultivated commercially only for a few hundred years. Before that, they were picked from wild vines or grown in gardens.

• Introduced to New Zealand in the first decade of the 20th century, kiwis were gradually improved and developed commercially to the point where they have become an extremely popular export crop. In the 1960s, California farmers imported Hayward kiwi vines from New Zealand and have been increasing production ever since.

• Though Europeans have known kiwis for about 100 years, they were considered to be an ornamental until their recent popularity. They are now grown commercially in southern Europe, Russia, Australia, Israel, and parts of South America, Africa and Asia.

Varieties and description

• The popular Hayward kiwi, egg-shaped and slightly flattened at both ends, is about 5 to 8 centimetres (2 to 3 inches) in length and 3.5 to 5 centimetres ($1\frac{1}{2}$ to 2 inches) in diametre. Kiwis are covered with a thin, fuzzy, brown skin. If cut crosswise, the kiwi's succulent, beautiful emerald green flesh is revealed. Tiny black edible seeds radiate out from its creamy-coloured heart.

• Kiwi fruits have a pleasant, exotic aroma, the texture of ripe melon and a delicious, sweet taste with a hint of acidity.

• There are a number of varieties, which differ slightly in shape, size and degree of sweetness. In China, where they are grown in the Jiangxi province, red, light yellow and golden-coloured kiwis can be found.

Availability

• Kiwis are available year round. New Zealand and South American kiwis are in season from spring to late fall, while California kiwis can be found from late fall to spring.

Selection

• Kiwis are harvested when firm, yet ripe, and they continue to get sweeter at room temperature, unlike most fruits, which must develop their sugars before harvest. Choose firm, unblemished fruit.

Storage

• Kiwis have a very long shelf life. Firm kiwis can be refrigerated for two or three weeks if necessary, then ripened for a few days until they yield to gentle pressure. To hasten ripening, put them in a paper bag with apples, pears or bananas. Once ripe, keep them away from these ethylene-producing fruits or they will overripen. Chill and use ripe kiwis or refrigerate for up to three days.

Preparation

• A simple way to eat a kiwi is to cut it in half lengthwise and scoop out the flesh with a spoon. For slices, discard a piece from each end, peel with a sharp knife and cut crosswise. Though the skin is edible, it is generally discarded.

• Cooked kiwi loses its bright colour. Add it toward the end of preparation.

• Actinidin is an enzyme found in kiwi that acts as a meat tenderizer; for improved texture and flavour, mash a few slices on your meat with a fork and let sit for at least 20 minutes before cooking. This enzyme both prevents gelatin from setting and makes milk products go off, unless the kiwi is lightly cooked first. If uncooked kiwi is used with milk products, serve soon after preparation.

Suggestions

• Kiwi juice is tasty alone or mixed with other juices. Try it in daiquiris or in cocktails. Don't overprocess kiwi juice or purée, as crushing the small black seeds releases a bitter flavour.

• For a healthy breakfast, chop a kiwi into cereal. (High in fibre, kiwis are a good natural laxative.)

• Garnishes, appetizers, poultry, meat or fish salads, entrées, fruit salad, glazed tarts, cake filling or topping, mousse and sorbet are just some of the uses for the versatile kiwi.

• Add sliced kiwis to orange sauce for duck, chicken, veal or fish fillets.

• For an elegant supper for two, melt a little butter in a frying pan, brown a large handful of slivered almonds and set aside. Add more butter and a minced clove of garlic to the pan. Dust 2 trout in seasoned flour, cook for a few minutes on each side and place in heated plates. Add a shot of brandy or wine to the pan and bring to the boil, seasoning to taste. Add two kiwis, sliced crosswise, reduce heat and cook for 1 minute each side. Arrange kiwi slices on top of the fish, drizzle on sauce and top with almonds. Serve with rice and cooked green vegetables. (Chicken breasts can be used instead of the fish.)

• Make an attractive pasta salad with kiwis for a buffet or a special lunch: Cook a package of your favorite small pasta shapes (bowties, rotini, penne, shells, etc.), rinse in cold water and drain well. Mix the following ingredients together and toss with pasta; 60 mL (4 TBSP) each olive oil and white vinegar, 15 mL (1 TBSP) Dijon mustard, 1 clove of minced garlic, and salt and pepper to taste. Add a little finely chopped fresh red or green basil, a handful of minced scallions, a finely chopped red or yellow pepper and a handful of black olives. Slice or dice 3 to 4 kiwis and mix well. If desired, add some cooked shrimp, crabmeat or lobster and adjust seasoning to taste.

• A delicious New Zealand dessert, kiwi Pavlova, is a mouth-watering treat; buy or make a plate-sized meringue shell, top with sweetened, vanilla flavoured whipped cream and decorate with thinly sliced kiwis. Kiwi shortcake is another favourite.

• Make an extra special lemon meringue pie by mixing 2 or 3 diced or sliced kiwis into the lemon filling.

Nutritional Value

Kiwi

(per 1 medium kiwi = 76 g = 2³/4 oz)

			RDA**	%
calories*	(kcal)	46		
protein*	(g)	0.8		
carbohydrates*	(g)	11.3		
fat*	(g)	0.3		
fibre	(g)	0.8	30	2.7
vitamin A	(IU)	133	3300	4
vitamin C	(mg)	75	60	125
thiamin	(mg)	0.02	0.8	2.5
riboflavin	(mg)	0.04	1.0	4
niacin	(mg)	0.4	14.4	2.8
sodium	(mg)	4	2500	0.2
potassium	(mg)	252	5000	5
calcium	(mg)	20	900	2.2
phosphorus	(mg)	31	900	3.4
magnesium	(mg)	23	250	2.2
iron	(mg)	0.31	14	2.2
water	(g)	63.1		

*RDA variable according to individual needs
**RDA recommended dietary allowance (average daily amount for a normal adult)

Pork tenderloin with kiwis

(serves 4)

Ingredients		
4	slices pork tenderloin, approximately 2 cm (3/4 inch) thick	
15 mL	butter	1 TBSP
15 mL	flour	1 TBSP
30 mL	white rum	2 TBSP
75 mL	pineapple juice	1/3 cup
190 mL	stock	3/4 cup
12	juniper berries, crushed	
	salt and pepper to taste	
6	kiwis, peeled and sliced	

Instructions

- In a heavy pan, cook the pork in butter over moderate heat, then transfer to a serving dish. Keep warm in the oven at 100 °C (200 °F).
- Drain off most of the fat from the pan, leaving about 15 mL (1 TBSP).
- Return the pan to medium heat and sprinkle on the flour, stirring to avoid lumps.
- Add the rum, stirring constantly, then slowly add the pineapple juice, stock and juniper berries. Bring to a boil and reduce until the sauce thickens. Lower heat and season to taste.
- Add sliced kiwis and heat for 1 minute.
- Arrange kiwi slices around the meat and drizzle sauce over the top.

Papaya

Hawaii ↑

LOW IN CALORIES
EXCELLENT SOURCE OF VITAMIN A AND C
GOOD SOURCE OF POTASSIUM AND MAGNESIUM
EXCELLENT FOR DIGESTION

➠ Origin

• The papaya, called pawpaw in many countries, is thought to have originated somewhere in southern Mexico, Central or South America. Its name derives from "ababai" — the Carib Indian name for the fruit. Papaya is the most important member of the Caricaceæ family, which contains such unusual fruits as the babaco, chamburo and chilguacán.

• Papaya trees spread easily in the wild. Long before the arrival of Europeans, this delectable fruit was enjoyed throughout the Latin American area. Spanish and Portuguese explorers introduced papaya to their Asian colonies from the 16th century onward, possibly through Africa.

• Preparations made from the fruit, seeds and leaves were traditionally used for beauty treatments and in indigenous herbal medicine for a variety of complaints, including digestive, liver and heart problems, dysentery, external wounds, arthritis and ulcers.

• Today, papayas are grown in dozens of countries, including the U.S. — primarily Hawaii and to a lesser extent Florida —, Mexico, Central and South America, the Caribbean islands, Africa, Australia, southern Asia and the Pacific islands.

➠ Varieties and description

• Depending on the variety, papayas can be globular, pear-shaped, football-shaped or cylindrical. They range from 10 to 50 centimetres (4 to 20 inches) in length. Generally, only the smaller pear-shaped varieties are grown for export — the most popular being the Hawaiian solo and sunrise and the Brazilian red Amazon. Its very thin, inedible, smooth skin starts off green and ripens to yellow or orange. Its flesh is yellow, orange, salmon pink or red and contains gray to black shiny seeds the size of large peppercorns in its hollow centre.

• Papaya flesh has a slightly firmer texture than that of a ripe, juicy melon. It has a mild, sweet, tropical flavour.

➠ Availability

• Papaya trees produce fruit year round. As there are so many producer countries, papayas are plentiful all year.

➠ Selection

• Papayas must be picked green, as ripe fruit is too fragile to travel to market. They turn yellow as they ripen over a period of a few days. It's hard to say which will be sweetest and most flavourful, as often a speckled and spotted skin covers the tastiest fruit. Even a small amount of mold won't hurt and can be cut away. Avoid completely green, rock hard papayas, fruit with large bruises and those that are over-soft.

➠ Storage

• Papayas should not be refrigerated until ripe, then they should be used within a few days — the sooner, the better the flavour. Keep them at room temperature until the skin turns from 3/4 to completely yellow and they yield slightly to very gentle pressure.

• In South America, a little trick is used to hasten ripening: Score the papaya skin lengthwise, as if dividing the fruit in quarters, but do not penetrate the flesh. Wipe clean and let ripen.

➠ Preparation

• The most popular way to enjoy a papaya is to cut it in half lengthwise, remove the seeds, sprinkle on a little fresh lemon or lime juice and scoop out the flesh with a spoon.

• Slices or chunks of papaya for garnishing hot entrées can be lightly grilled, baked or sautéed; more than a few minutes of cooking may diminish its delicate flavour, but not its firm texture.

• Papayas contain the enzyme papain, which is an excellent meat tenderizer because it digests protein. Papain prevents gelatin from setting and can curdle milk products after a while. Serve papaya combined with dairy foods (such as a papaya shake or papaya halves filled with ice cream) soon after preparation, making only enough for immediate use.

• Papaya seeds can be dried and used for seasoning, as you would use crushed black pepper.

• In many tropical countries, the larger varieties of papaya are used as a vegetable when still green and hard. They are baked or cooked like squash, grated and tossed with salad dressing, or made into chutney or pickles.

Suggestions

• For a healthy breakfast, try papaya chunks in yogurt, or a papaya shake: In a blender, mix a papaya, a banana, a spoon of honey, a squeeze of fresh lemon or lime and 2.5 mL (1/2 tsp) vanilla with a small glass of milk, buttermilk or yogurt. A scoop of ice cream makes it a real treat!

• For an elegant salad, combine a diced papaya with various salad greens and toss in a spicy mustard vinaigrette.

• Puréed papaya is ideal to add to marinades, as it is an excellent meat tenderizer. For good flavour and texture, save the peel and lay it flesh side down or apply thin slices of papaya to all types of meat and poultry, letting it stand for about 30 minutes before cooking.

• Try fruit kebabs with papaya and pineapple chunks, alternated with tangerine segments, grapes, strawberries, etc. Sprinkle on a little liqueur or gingery syrup if desired. Papaya chunks also make appetizing tidbits combined with ham or prosciutto and cheese, skewered with toothpicks.

• Grilled or barbecued fish or chicken is well complimented by a tangy papaya salsa: Dice a papaya into small cubes, finely mince a half to a whole mild or medium chili pepper, mince 1 garlic clove and 1 small onion, finely sliver 1/2 a red pepper, add 30 mL (2 TBSP) olive oil, the juice of 1 or 2 limes, 15 mL (1 TBSP) finely minced cilantro or parsley and season to taste. Mix ingredients and refrigerate for a few hours, stirring now and then.

• Cut papayas in rings crosswise, about 3 centimetres (1 inch) thick. Peel off skin and discard seeds. Serve 1 to 4 rings per person (depending if it's an appetizer or a salad meal), filling the rings with a salad of chicken, turkey, ham, cottage cheese, shrimp or crab. Papaya halves are also good for stuffing with a variety of sweet or savoury fillings.

Brazil →

Papaya carrot cake

Ingredients		
250 mL	sugar	1 cup
250 mL	vegetable oil	1 cup
3	eggs	
2.5 mL	salt	1/2 tsp
6 mL	baking soda	1 1/4 tsp
6 mL	baking powder	1 1/4 tsp
5 mL	cinnamon	1 tsp
330 mL	all-purpose flour	1 1/3 cups
250 mL	grated carrots	1 cup
250 mL	puréed papaya	1 cup
125 mL	diced pineapple	1/2 cup
60 mL	walnuts, chopped	1/4 cup
Icing		
500 mL	cream cheese	2 cups
500 mL	icing sugar	2 cups
15 mL	butter, softened	1 TBSP
handful of candied orange zest for garnish		

Instructions

- Beat together sugar, oil and eggs.
- Sift together salt, baking soda, baking powder, cinnamon and flour. Gradually add dry mixture to egg mixture.
- Beat for 3 to 4 minutes and add the carrots, fruit and nuts, beating until mixture is completely smooth.
- Pour batter into greased tube or bunt pan and bake 45 to 60 minutes in a preheated oven at 180 °C (350 °F). Cool on a rack.
- Beat cream cheese, icing sugar and butter until smooth and creamy, ice the cake, then decorate with candied orange zest.

Nutritional Value
Papaya
(per 1 medium papaya = 304 g = 10 3/4 oz)

			RDA**	%
calories*	(kcal)	117		
protein*	(g)	1.9		
carbohydrates*	(g)	29.8		
fat*	(g)	0.4		
fibre	(g)	2.4	30	8
vitamin A	(IU)	6122	3300	185.5
vitamin C	(mg)	188	60	313.3
thiamin	(mg)	0.08	0.8	10
riboflavin	(mg)	0.1	1.0	10
niacin	(mg)	1	14.4	6.9
sodium	(mg)	8	2500	0.3
potassium	(mg)	780	5000	15.6
calcium	(mg)	72	900	8
phosphorus	(mg)	16	900	1.8
magnesium	(mg)	31	250	12.4
iron	(mg)	0.3	14	2.1
water	(g)	270		

*RDA variable according to individual needs
**RDA recommended dietary allowance (average daily amount for a normal adult)

Carambola/star fruit

Origin

• The exotic carambola *(Averrhoa carambola)* or star fruit, and its close relative the cucumber tree fruit *(Averrhoa bilimbi),* are the only edible fruits in the Oxalidaceæ (wood sorrel) family. While carambolas can be sweet or tart, the similar-looking, though smaller cucumber tree fruit, is always sour and not often seen outside Southeast Asia.

• Carambolas originated in Southeast Asia or India thousands of years ago and have long been a popular fruit in China, India and throughout tropical Asia. Taken to Europe by returning travellers in the late 1700s, they became a popular ornamental in the hot houses of the rich.

• Chinese immigrants or merchant traders probably were the ones to establish the carambola in Hawaii, from where it was introduced to Florida about 75 years ago. Florida now is the principal source for the star fruit consumed in the U.S. Among some of the biggest producer-exporter countries are Brazil, New Zealand, Australia and Taiwan.

Varieties and description

• Averaging 7 to 15 centimetres (3 to 6 inches) in length and 5 to 8 centimetres (2 to 3 inches) in diametre, this unique fruit has five sharply angled ribs and looks like an elongated five-point star. It has a very thin, edible, lime green skin that ripens to bright yellow, with a waxy sheen to it.

• The pale yellow, juicy flesh, which contains a few, small, flat seeds, can be crisp or soft, and has a distinctly fruity tropical flavour. Some varieties of carambola are very sweet, others are slightly tart.

Availability

• Carambola trees can produce up to three crops a year. They are available from July to May; Florida carambolas are available from August to February, Brazilian carambolas from January to March and July to October, while those from New Zealand and Australia are found from October to December.

Selection

• Choose firm, glossy-skinned, unbruised carambolas. Browning on the tips of the ridges is a sign of ripeness.

Storage

• Leave at room temperature until most traces of green have disappeared and the fruit has turned golden yellow, possibly with traces of brown on the edges. Ripe carambolas have a lovely, fragrant aroma.

VERY LOW IN CALORIES
GOOD SOURCE OF VITAMINS A AND C
CONTAINS MINERALS

• If not too ripe, they can be kept refrigerated in a paper bag for about a week. Once ripe, eat them within a day or two.

Preparation

• The entire fruit is edible, except for the seeds. A sweet carambola is a delicious, refreshing, low-calorie snack eaten as is, after washing. Peel off the skin if you prefer.

• Slice thick or thin star shapes crosswise to use as a beautiful garnish. You may want to pare off any brown edges for a more elegant look.

• Carambolas should be cooked only lightly. The tart varieties taste sweeter once heated.

Suggestions

• Cut carambolas in half crosswise and juice them on a

citrus juicer, for an excellent drink, alone, in tropical cocktails or mixed with other fruit juices. This juice can also be used to make sorbet.

• Lightly sauté stars and serve with roast poultry, meat or fish.

• Carambola stars are excellent in sweet and sour dishes.

• Use sliced stars to garnish drinks or punch bowls, seafood or cottage cheese salads, buffet platters, cheese

platters, baked hams or desserts. (Glaze the stars in a light syrup when using them to decorate pies or cakes.)

- In Asian cuisines, carambolas are used in curries and for pickles and chutneys. Try this simple western-style curried salad: Combine 500 mL (2 cups) of cooked rice, 250 mL (1 cup) cooked, diced chicken breasts, 1 small minced onion, 2 minced celery stalks, chopped green and red pepper (half each), and 1 diced carambola. Mix together 45 mL (3 TBSP) each mayonnaise, sour cream, and lemon juice, and 2.5 mL (1/2 tsp) curry powder. Mix everything together, season to taste and refrigerate. Serve garnished with stars.
- Spread sliced stars on a plate and place in the freezer. When solid, put them in a sealed bag or container and keep them frozen for later use as instant popsicles or to garnish frozen desserts.

Nutritional Value

Carambola

(per 1 medium carambola = 127 g = 4 1/2 oz)

			RDA**	%
calories*	(kcal)	42		
protein*	(g)	0.7		
carbohydrates*	(g)	9.9		
fat*	(g)	0.4		
fibre	(g)	1.2	30	4
vitamin A	(IU)	626	3300	18.9
vitamin C	(mg)	27	60	45
thiamin	(mg)	0.04	0.8	5
riboflavin	(mg)	0.03	1.0	3
niacin	(mg)	0.5	14.4	3.5
sodium	(mg)	2	2500	0.08
potassium	(mg)	207	5000	4.1
calcium	(mg)	6	900	0.7
phosphorus	(mg)	20	900	2.2
magnesium	(mg)	12	250	4.8
iron	(mg)	0.33	14	2.4
water	(g)	115.5		

*RDA variable according to individual needs
**RDA recommended dietary allowance (average daily amount for a normal adult)

Sautéed carambolas with cardamon

(serves 4 to 6)

Ingredients

15 mL	butter	1 TBSP
3	carambolas, in 7 mm (1/4 inch) slices	
1/2	each green, red and yellow pepper, cut in strips	
60 mL	chicken stock	1/4 cup
45 mL	honey	3 TBSP
15 mL	crushed cardamon seeds	1 TBSP
	salt and pepper to taste	

Instructions

- Melt butter; sauté carambola slices and peppers for about 4 minutes over medium heat.
- Add chicken stock, honey and cardamon; cook for a minute or two.
- Season to taste and serve as a side dish with fish, poultry or meat.

Artichoke

Low in calories
High in fibre
Good source of minerals

Origin

• Originally from the Mediterranean area, the artichoke is an edible flower bud of a thistle, picked before it blooms. (If the bud is left on the plant, it develops into a large, beautiful purple, spikey flower.) A member of the Compositeæ family, which also includes lettuce, endive and cardoon — another edible thistle and the ancestor of the artichoke —, the artichoke's name derives from the Ligurian word "cocal", meaning pine cone, which it resembles.

• Artichokes marinated in vinegar were an expensive delicacy at the height of the Roman empire. They were next heard of when several noted Italian gardeners began to cultivate them in the 15th century. Catherine de Medici adored them and introduced them to the French royal court, where they became very popular. The private physician of Louis XIII claimed that artichokes warm the blood and inflame the senses. Artichokes were once prescribed for the rich to treat insomnia, liver and kidney problems and as a beauty treatment.

• With time, their popularity spread to Spain. They were introduced to many Spanish and French colonies. French settlers brought them to Louisianna, Spaniards to California. With the arrival of Italian immigrant farmers in California in the 1920s, the artichoke began to be grown in increasing quantities.

• Today, California is one of the world's largest producers, along with France, Italy, Spain, Egypt and Israel.

Varieties and description

• The artichoke is a soft sage green colour, sometimes tinged with purple, and ranges from globular to cone-shaped. It has an edible stem connected to its tender, meaty, delicious, cup-shaped heart. Nestled in the heart is the inedible, thistle-like "choke", which is easily removed after the artichoke has been cooked. Surrounding the heart and choke are the artichoke's pointed leaves, edible at the base and sometimes sharp at the tips. They gradually grow larger up to the outermost leaves, which are the toughest. Artichokes have an unusual, delicate flavour and cannot be eaten raw.

• Mature artichokes come in three sizes, which grow on different parts of the plant. Small or baby artichokes, no bigger than 5 centimetres (2 inches) in diametre, often have no choke at the centre. The medium size averages 225 to 285 grams (8 to 10 ounces), while the large or jumbo size averages 425 to 570 grams (15 to 20 ounces).

Availability

• Artichokes are available year round. The seasonal peaks in California are from March to May and in October. Baby artichokes are usually found in the spring.

Selection

• Choose fresh-looking artichokes that feel heavy for their size. Look for tightly packed leaves.

• "Winter-kissed" artichokes, available in the fall and winter, are considered more tender and flavourful by some connoisseurs. They can be darker, have bronze-tipped leaves, or even a whitish tint from exposure to low temperatures.

Storage

• Unwashed artichokes can be kept in the refrigerator for about a week. Sprinkle a few drops of water on them and seal them in an airtight bag.

Preparation

• First, wash the artichokes well, trim the stem to about 2.5 centimetres (1 inch) and snap off the small bottom leaves. Some chefs recommend snipping off the tips of the leaves or cutting off the top 1/4 to 1/3 of the artichoke; this is not really necessary unless you are preparing a fancy meal or are making baby artichokes to be eaten whole. If you do cut them, use a stainless steel knife and rub lemon on cut surfaces.

• As each artichoke is prepared, drop it in a large pot (not aluminum or cast iron) half-filled with water. Add the juice of a lemon or a little vinegar, 15 mL (1 TBSP) olive oil, 1 or 2 whole garlic cloves, a quartered onion, 1 or 2 bay leaves, a little salt and a few whole pepper corns. Bring to a boil, cover, then simmer gently for 25 to 45 minutes, depending on size and quantity. It is preferable to overcook artichokes rather than to undercook them. When they are done, a central leaf will pull out easily. Remove and drain upside down for a few minutes before serving. Baby artichokes will cook in 10 to 15 minutes.

• Alternatively, steam for about an hour, upside down, or cook in a pressure cooker for 8 to 10 minutes.

Fiddleheads

VERY LOW IN CALORIES
EXCELLENT SOURCE OF VITAMINS A, COMPLEX B, POTASSIUM AND IRON

Origin

• Fiddleheads are an exotic vegetable available only briefly each year, at the beginning of the growing season. This delicacy, a member of the Polypodiaceæ family, was known to the Native North Americans long before the arrival of Europeans. They are found in other parts of the world as well, but their origins are not known.

• Fiddleheads are the coiled tips of the early spring shoots of ferns, picked when the fern is only a few inches high. Once sprouted, they must be harvested within a few days, or the tips will begin to uncurl into fronds and they will no longer be edible. All ferns look like fiddleheads when they begin to grow in the spring; although the tips of several other species have traditionally been locally harvested and eaten when young and tender, it is only the tips of the ostrich fern and the cinnamon fern that are recommended as completely safe to eat. Generally, only ostrich fern tips are marketed as fiddleheads.

• Fiddleheads are harvested mainly in New Brunswick and Quebec, though they can be found in other Canadian Atlantic provinces, in Maine and Vermont, and in other northeastern states. They are still gathered by hand from their natural habitat. Research is under way to try to develop them as a commercial farm crop.

Description

• Fiddleheads are aptly named for their appearance. Tightly furled, they look just like the head of a violin. These jade green coils, which grow covered in brown papery scales, should be 2 to 4 centimetres (1 to 1 1/2 inches) in diametre and have only a short piece of stem extending beyond the coil.

• The taste, somewhat like wild asparagus, has also been compared to artichoke hearts and wild mushrooms.

Availability

• Fiddleheads are harvested over a two week period, anywhere from the middle of April to the beginning of July. They are available to the public for a month or two, as the season in each growing area differs slightly.

Selection

• Choose the most fresh, firm and tightly curled fiddleheads. Some packaged fiddleheads have had their scales removed, while others need a little cleaning before cooking.

Storage

• Keep fiddleheads refrigerated, well wrapped with a paper towel to absorb excess moisture, in a plastic bag. For the best flavour and texture, use them within a day or two of purchase. Wash only when ready to use.

• Fiddleheads can be frozen for later use: Blanch them in boiling water for a minute or two, plunge in cold water, dry thoroughly and freeze in well-sealed bags or containers.

Preparation

• Gently rub fiddleheads between your hands to remove the brown scales, or shake them in a paper bag. Carefully wash, drain dry, then trim off the stems.

• Fiddleheads taste best when added to boiling, lightly salted water and cooked for five to seven minutes, though they can also be steamed or sautéed. Be careful not to undercook or overcook them.

• If you want to serve them several days after purchase, it's best to cook them first, plunge them in cold water and drain well. Wrap them and keep them refrigerated, then either reheat them in a little butter or serve cold in a vinaigrette.

Suggestions

• For a delicious side dish, sauté cooked fiddleheads for a minute in a little butter or margarine. A squeeze of lemon or a little white wine vinegar and a little salt and pepper bring out the flavour wonderfully. Use any leftovers in an omelet.

• Serve fiddleheads with hollandaise sauce or au gratin in Mornay or bechamel sauce — perfect with grilled fish, chicken or steak.

• Fiddleheads are excellent in soup, especially clear chicken broth.

• Create a gourmet salad with cooked fiddleheads and salad greens. Toss with a traditional vinaigrette, or an oriental dressing: mix 30 to 45 mL (2 to 3 TBSP) sesame oil, 30 mL (2 TBSP) black Chinese vinegar or soy sauce, a little grated fresh ginger and a few crushed fennel seeds. Add a pinch of salt if necessary. Add well drained canned baby corn, minced scallions and finely slivered coloured peppers.

• Use fiddleheads instead of spinach for an exotic version of eggs Benedicts or Florentine.

• Boil fiddleheads until tender, then melt a little butter or margarine over them. Lightly season them, place in an oven proof dish and sprinkle generously with Parmesan cheese. Broil for 1 minute, until the cheese browns nicely.

Nutritional Value

Fiddleheads

(per 150 g = 5¼ oz)

			RDA**	%
calories*	(kcal)	30		
protein*	(g)	6.4		
carbohydrates*	(g)	5		
fat*	(g)	0.7		
fibre	(g)	1.7	30	5.6
vitamin A	(IU)	3263	3300	98.9
vitamin C	(mg)	30	60	50
thiamin	(mg)	0.03	0.8	3.4
riboflavin	(mg)	0.38	1.0	38
niacin	(mg)	7	14.4	48.6
sodium	(mg)	2	2500	0.08
potassium	(mg)	332	5000	6.6
calcium	(mg)	8	900	0.9
phosphorus	(mg)	157	900	17.4
magnesium	(mg)	51	250	.20
iron	(mg)	1.7	14	12
water	(g)	93		

*RDA variable according to individual needs

**RDA recommended dietary allowance (average daily amount for a normal adult)

Fiddleheads in mushroom sauce

(serves 4-6)

Ingredients

500 g	fiddleheads	1 lb
60 mL	butter	4 TBSP
24	mushrooms, quartered	
5 mL	orange zest	1 tsp
	cumin, ginger, salt and pepper to taste	

Instructions

• Clean, then steam or boil the fiddleheads until tender.

• In a heavy pan, melt butter and add mushrooms and fiddleheads.

• Add the zest and spices and sauté until the mushrooms are done.

Avocado

Good source of vitamin A
Excellent source of vitamin B complex, potassium, magnesium and iron
Contains no cholesterol

Origin

• The avocado belongs to the Lauraceæ (laurel) family, which consists mainly of aromatic plants, such as the laurel, sassafras, cinnamon and camphor. Originally from the tropical areas of Central and South America, the avocado had spread as far north as Mexico more than 2,000 years ago, but was not found in the Caribbean islands until brought there by early European explorers. The Aztec name, "ahuacatl", from which the word avocado derives, meant "butter from the wood". Other names include alligator pear, butter pear and vegetable pear.

• Attempts were made to introduce avocados to Florida and California in the 19th century, but it wasn't until early this century that successful commercial development began. It is only in recent years, however, that the admirable avocado, once an exotic luxury from Latin America, has gained enormously in popularity.

• Though most of our avocados come from California and Florida, they are grown extensively in Mexico, Central and South America, Africa, Spain and Israel.

Varieties and description

• Avocados differ greatly in size, shape, texture and size of seed, as well as in the colour, thickness and texture of the skin. The smooth, easily digestible flesh is yellowish green, at times quite pale. In some varieties, the flesh adheres tightly to the seed, while in others, the seed is loose and can be heard rattling around if the avocado is shaken.

• Several hundred varieties exist, which are hybrids bred from one or more of the three main types of avocado: Mexican, Guatemalan and West Indian. The Mexican one is small and has smooth, purple to black skin; the Guatemalan is medium-sized and has leathery, lightly pebbled, green, purple or black skin; and the West Indian type is large and has smooth, green or reddish-purple skin.

• Of the six well known varieties grown in California, Hass avocados account for more than 80 per cent of the crop. The oval, creamy-fleshed **Hass** avocado is dark green (to black when ripe) with a heavily pebbled skin of medium thickness and a small seed. The popular, medium-sized, pear-shaped **Fuerte** has dark green, lightly pebbled, thin skin. Other notable hybrids include the **Bacon** and **Zutano** varieties, which are both green and pear-shaped. The former has a thicker skin, while the latter is slightly more elongated; the **Pinkerton,** is also pear-shaped and green-skinned; and the **Reed,** a rounder-shaped avocado has a thick, heavily pebbled skin.

• The unusual tiny cocktail avocados, which are really a freak of nature, are sometimes available. These are avocados with no seed, usually of the Fuerte variety, and are only rarely harvested by growers.

• Florida varieties are mostly hybrids of the West Indian type, often crossbred with the Guatemalan avocado. They are characteristically very large, with thin, smooth, green skin. Most Florida avocados have a much lower oil content than Californian varieties and, therefore a less rich and creamy flavour and texture.

• Avocados are a delicious and nutrient-dense fruit. They contain 17 vitamins and minerals and, like all fruits and vegetables, contain absolutely no cholesterol.

Availability

• Avocados are most plentiful in summer, though they can be found year round, as there are so many producer countries. Hass avocados have a long growing season and are available from January to November. Greenskin varieties — Fuerte, Pinkerton, Bacon and Zutano — generally are available from November through May. Reed avocados are in season from July to September.

• Florida avocados are most plentiful in summer and fall.

Selection

• Avocados are harvested when mature, but not quite ripe enough to eat. Generally, they must be bought a few days before they are needed and ripened at home. They are at their peak when they yield slightly to gentle pressure. Be careful not to bruise an avocado by squeezing too hard.

Storage

• Never refrigerate an avocado before it has fully ripened. To speed up ripening, wrap in newspaper or a paper bag. Once ripened and refrigerated, use within a week.

1
2
2
3
3
1. Florida jumbo
2. Florida
3. Hass

Rapini

VERY LOW IN CALORIES, HIGH IN FIBRE
EXCELLENT SOURCE OF VITAMINS A AND C
VERY GOOD SOURCE OF MINERALS

Origin

• Rapini, also known as Italian broccoli, broccoli-rabe or broccoletti di rape, is closely related to broccoli and possibly is its ancestor. A member of the Cruciferæ family, rapini is part of the important genus Brassica, which includes many types of cabbage, cauliflower, Brussels sprouts, kohlrabi and kale. Like broccoli, rapini is originally from the Mediterranean area and has been popular in Italy for hundreds of years.

• Italian settlers to the United States introduced rapini during the 1920s.

• It is only recently that other communities are discovering and appreciating it. Today, it is grown in California, Arizona, New Jersey, Quebec and Ontario, among other places.

Varieties and description

• Rapini is medium to dark green in colour, and resembles broccoli, except that it has very slender stalks, many of which end in leaves. Instead of one large head, some of the stalks end in small clusters of tiny green buds, at times sprouting small yellow flowers. The entire plant is edible — stems, leaves, heads and flowers.

• Rapini has a very pleasant, slightly bitter flavour. The leaves have a stronger taste than the rest.

Availability

• Rapini can be found year round; the season in California and Arizona runs from September to April; in New Jersey from May to September; and in Canada from April to August.

Selection

• Look for freshly coloured, firm rapini, avoiding wilted or yellowed leaves or stems.

Storage

• Keep unwashed rapini wrapped and refrigerated and use within a few days of purchase — the fresher, the better the flavour.

Preparation

• Wash well, then trim the bottoms off the stalks. Separate the stalks, place them in a very small amount of boiling water and cook for a minute, then add the rest and simmer for only 2 to 4 minutes, being careful not to overcook it. The texture should be barely tender and the leaves should retain their colour.

• Rapini can be sautéed, braised or microwaved as well, but it sometimes tastes bitter when steamed.

Suggestions

• Substitute rapini for broccoli in most dishes. Try it in a cheese sauce au gratin, creamed, or simply with a squeeze of lemon and a dab of butter.

• Serve cold or hot rapini dressed with a little olive oil, lemon or white wine vinegar, a touch of garlic and salt and pepper to taste. A few chopped anchovies and scallions are delicious in this dish.

• Rapini is good in soups and stews.

• Rapini is excellent in a simple pasta dish with olive oil, garlic and Parmesan cheese. Add some finely chopped Italian sausage for extra flavour.

Rapini frittata

(serves 4)

Ingredients		
1	bunch rapini	
15 mL	olive oil	1 TBSP
1	medium onion, finely sliced	
2	garlic cloves, finely minced	
1	red pepper, slivered	
6	eggs	
	salt and pepper to taste	
75 mL	grated cheese	1/3 cup

Instructions

- Chop rapini coarsely and blanch for 2 or 3 minutes. Drain well.
- In a large, heavy frying pan, heat oil on medium. Add onions and garlic and stir, then add red pepper after 2 minutes. When onions are tender, add the rapini.
- Beat eggs lightly with seasonings, pour over vegetables, then sprinkle on the cheese. Cook on medium-low until the eggs are nearly set, finishing under the broiler for the last 2 minutes.
- Serve with salad.

Nutritional Value

Rapini

(per 100 g = 3 1/2 oz)

			RDA**	%
calories*	(kcal)	32		
protein*	(g)	3.6		
carbohydrates*	(g)	5.9		
fat*	(g)			
fibre	(g)	1.5	30	5
vitamin A	(IU)	2500	3300	75.7
vitamin C	(mg)	113	60	188
thiamin	(mg)	0.1	0.8	12.5
riboflavin	(mg)	0.23	1.0	23
niacin	(mg)	0.9	14.4	6.2
sodium	(mg)	15	2500	0.6
potassium	(mg)	382	5000	7.6
calcium	(mg)	103	900	11.4
phosphorus	(mg)	78	900	8.6
magnesium	(mg)		250	
iron	(mg)	1.1	14	7.9
water	(g)	89		

*RDA variable according to individual needs
**RDA recommended dietary allowance (average daily amount for a normal adult)

Rapini quiche

(serves 4)

Ingredients		
30 mL	grated Parmesan cheese	2 TBSP
1-2	bunches rapini, chopped coarsely	
1	medium onion, minced	
15 mL	olive oil or butter	1 TBSP
200 g	grated Gruyère or cheddar cheese	7 oz
2	eggs	
75 mL	sour cream	1/3 cup
60 mL	light cream	1/4 cup
2 mL	freshly grated nutmeg	1/4 tsp
	salt, pepper to taste	

Instructions

- Sprinkle the bottom of an **uncooked pastry shell** with Parmesan cheese. Cook rapini 2 to 3 minutes, drain well and place in pie shell.
- Sauté onion in oil until tender, but not browned. Add to pie shell. Sprinkle grated cheese evenly over rapini and onions. Whisk together the rest of the ingredients and carefully pour into the shell.
- Preheat oven to 200 °C (400 °F) and bake until done, about 40 minutes.

Asparagus

Origin

• Asparagus is a member of the Liliaceæ family, which includes lilies, gladioli and many other flowers, as well as a number of vegetables such as leeks, onions and shallots. The asparagus we enjoy today has been developed from a wild variety, indigenous to an area stretching from the eastern Mediterranean to Asia Minor. Its name comes from the Greek word "aspharagos", meaning sprout, and its American nickname, dating back to colonial times, is sparrowgrass.

• Asparagus was highly prized in ancient civilizations, more for medicinal purposes than as a food, until the Romans began cultivating it and spreading it throughout their empire. Early settlers to North America brought asparagus with them, though it wasn't planted on a large scale until the later half of the 19th century.

• Today, asparagus is cultivated extensively in the United States and to a lesser extent in Canada. It is also grown in Mexico, Spain, France, Italy, Holland and other European countries, Australia, New Zealand and in many South American countries.

Varieties and description

• Asparagus can vary in thickness and colour. They are ivory-coloured, violet or fresh dark green — the most popular type in North America. White or blanched asparagus, especially prized in Europe, is grown with earth heaped up around it, to keep the sunlight off.

• Each spear, averaging the thickness of a finger, is about 20 centimetres (8 inches) long. The tender tips are covered with tiny buds that would develop into ferny leaves if left unharvested.

Availability

• Asparagus is available nearly all year, as there are so many producer countries. It is most plentiful from March to June.

Selection

• Look for firm, fresh, crisp stalks and tightly closed, compact tips. Thick or thin, the spears taste alike, but it's best to choose a uniform size, so they will cook in the same time.

Storage

• Many growers recommend that asparagus should be well washed in lukewarm water to remove any sand that may be clinging to the tips, dried, wrapped in a slightly dampened paper towel, placed in a plastic bag and refrigerated. Some people cut a thin slice off the bottom and stand them in a little cold water in the refrigerator. Asparagus is highly perishable and should be used within a day or two of purchase.

VERY LOW IN CALORIES
GOOD SOURCE OF B AND C VITAMINS
EXCELLENT SOURCE OF IRON

Preparation

• Break off the woody bottom of each spear, retaining as much of the stalk as possible. Wash well, if you haven't yet done so. Some chefs recommend peeling the stalks, but this is rather wasteful.

• The traditional way to cook asparagus is in a special tall pot or a double boiler, with the top half turned upside down to accommodate its height. Tie the asparagus in bundles of 6 spears and stand them up in about 5 centimetres (2 inches) of lightly salted boiling water. Cover and cook 5 to 10 minutes, until the bottom is just tender, then drain immediately.

- Asparagus can also be cooked loose in a frying pan in no more than 2.5 centimetres (1 inch) of boiling water. Cover and simmer for a few minutes. Asparagus should never be overcooked or it will lose its fresh green colour and excellent texture.
- It must remain crisp, yet tender. If served cold, plunge immediately into cold water, then drain well.
- Asparagus can be also cut and stir-fried for a few minutes in hot olive oil or microwaved in a little water; cooking time depends on thickness of spears and quantity.

Suggestions

- One of the simplest and tastiest ways to enjoy fresh asparagus is with a little melted salted butter, adding a squeeze of lemon juice if desired.
- Serve asparagus hot in a great variety of ways: sprinkled with a little Parmesan cheese and garlic butter and lightly browned under the broiler; rolled in bread-crumbs, herbs and minced scallions and browned in a little hot olive oil or butter, seasoned to taste; lightly coated with hot lemon or orange sauce; sprinkled with butter and toasted slivered almonds; or in a cheese sauce, au gratin.
- A classic dish is asparagus with hollandaise sauce: In a double boiler, over hot though not boiling water, add 2 large or 3 small beaten egg yolks and 15 mL (1 TBSP) lemon juice. Using a whisk, constantly stir in 120 mL (8 TBSP) butter, a little at a time, until the sauce has thickened properly. Whisk in salt and pepper to taste and serve immediately over hot asparagus.

Nutritional Value

Asparagus

(per 100 g = 3½ oz)

			RDA**	%
calories*	(kcal)	21		
protein*	(g)	2.5		
carbohydrates*	(g)	3.6		
fat*	(g)	0.3		
fibre	(g)	0.9	30	3
vitamin A	(IU)	633	3300	19
vitamin C	(mg)	20	60	33.3
thiamin	(mg)	0.15	0.8	18.8
riboflavin	(mg)	0.18	1.0	18
niacin	(mg)	1.8	14.4	12.9
sodium	(mg)	5	2500	0.2
potassium	(mg)	221	5000	4.4
calcium	(mg)	16	900	1.8
phosphorus	(mg)	59	900	6.6
magnesium	(mg)		250	
iron	(mg)	1.4	14	10
water	(g)	92.7		

*RDA variable according to individual needs
**RDA recommended dietary allowance (average daily amount for a normal adult)

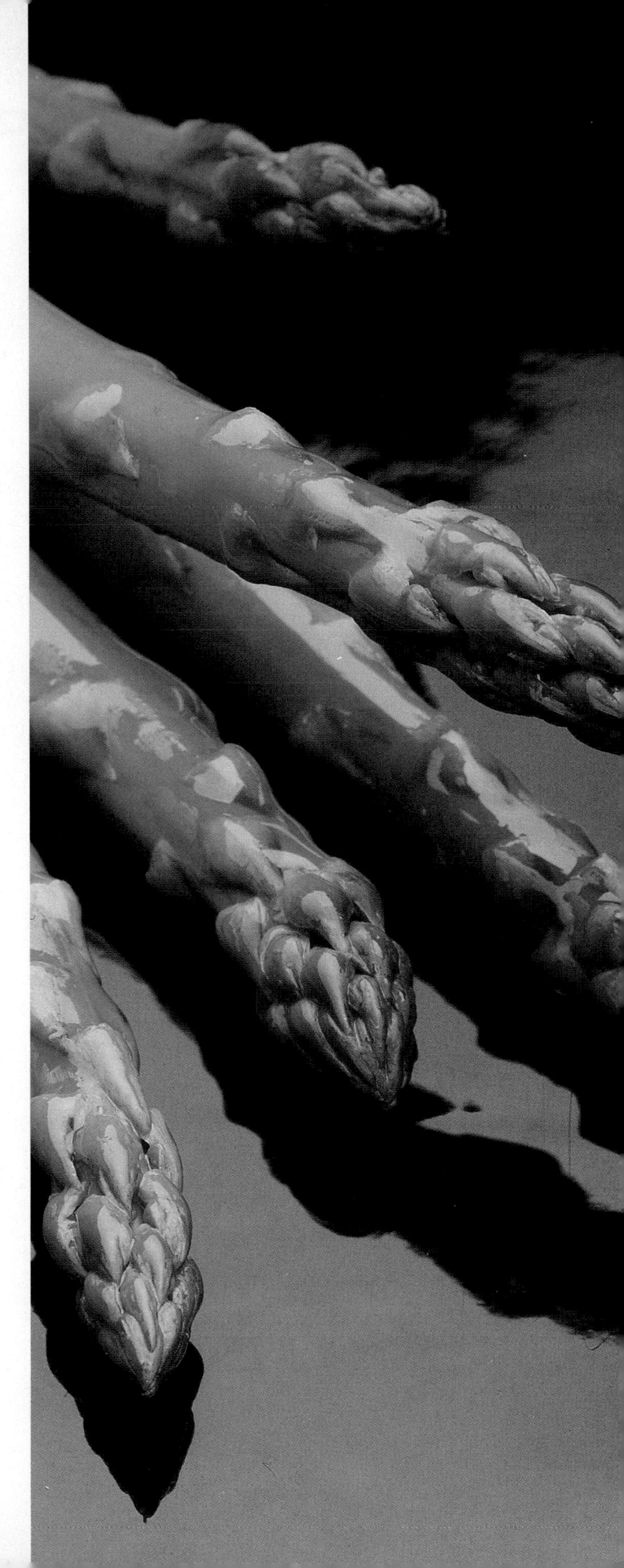

Escargot and asparagus in puff pastry

(serves 4)

Ingredients		
30 mL	butter	2 TBSP
24	Bourgogne escargots	
60 mL	white wine	1/4 cup
1	tomato, peeled and finely chopped	
1	garlic clove, minced	
	thyme and rosemary to taste	
375 mL	35 % cream	1 1/2 cups
24	medium asparagus, peeled and half-cooked	
	salt and pepper	
4	puff pastry shells	

Instructions

- Melt half the butter in a pot and sauté the escargot for 1 minute over medium heat.
- Remove from heat, drain and set aside.
- Put the pot back on the stove and deglaze it with the white wine.
- Add tomato, garlic and herbs.
- Reduce liquid to about 15 mL (1 TBSP).
- Add the cream and simmer. When the sauce begins to thicken, add the asparagus and escargot.
- Simmer 1 more minute. Add remaining butter.
- Stir slowly, season to taste and fill the preheated pastry shells.
- Serve as a first course.

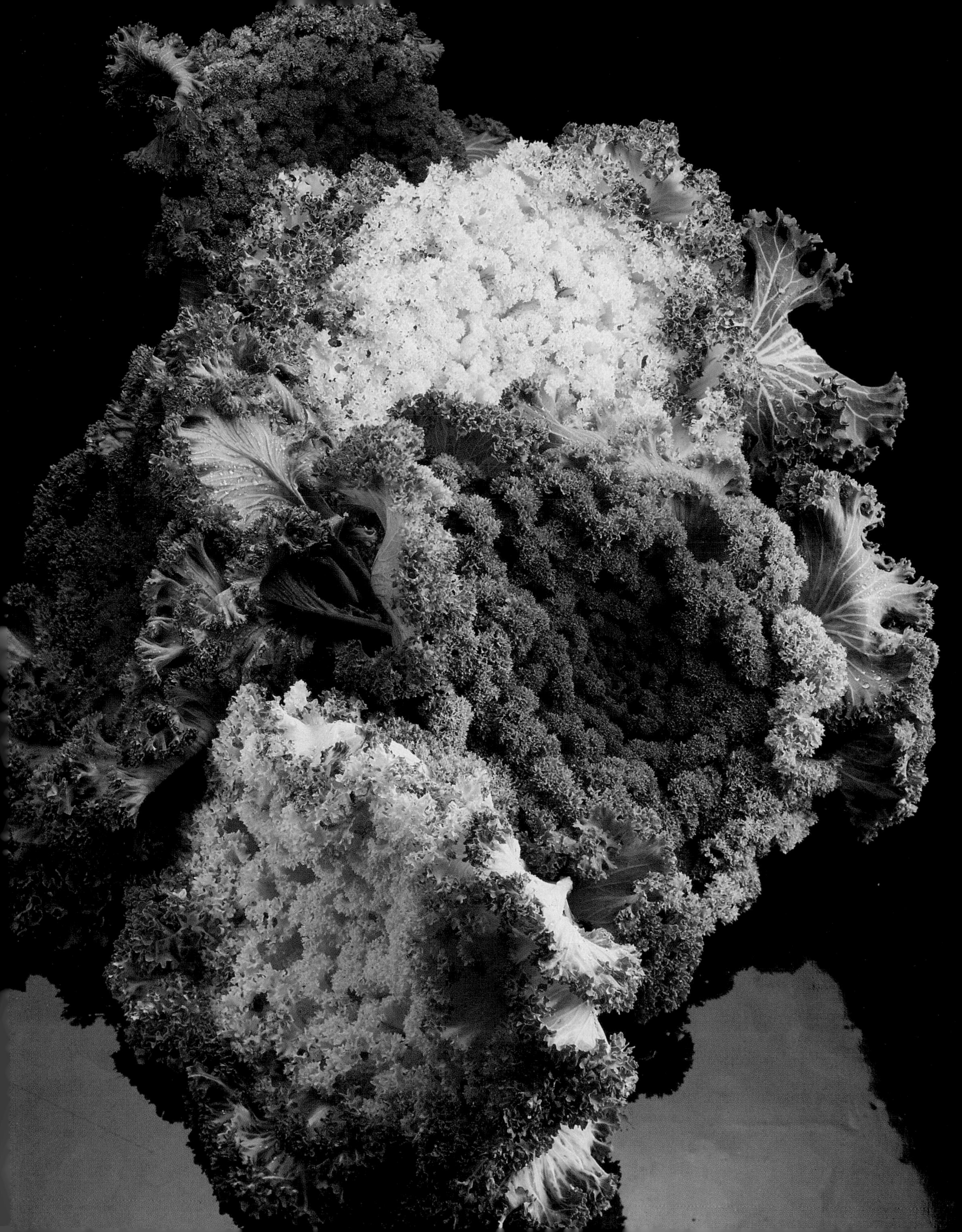

Savoy salad

EXCELLENT SOURCE OF VITAMINS A AND C
VERY GOOD SOURCE OF MINERALS
PARTICULARLY IRON AND CALCIUM
LOW IN CALORIES

Origin

• Sometimes called flowering cole, flowering kale or savoy lettuce, savoy salad is a recently developed vegetable that is closely related to cabbage and to kale, but has its own unique flavour and ornamental colour. A member of the Crucifer family, which also includes broccoli, kohlrabi and cauliflower, savoy salad should not be confused with yellow or green savoy cabbage. In fact, the word "savoy" means curly leafed.

• Savoy salad was discovered growing in a flower garden in Sweden by John Moore, a vegetable grower from California, who was so excited by its attractive appearance that he brought seeds home with him and developed a better tasting plant from them.

Varieties and description

• Savoy salad is so beautiful that it looks more like a flower than a vegetable. It has loosely packed crinkly leaves that are either purple, pink or cream-coloured, and are edged in green or white, while the outer leaves are green. It has a crispy texture like cabbage (only softer), with a mellow taste — slightly reminiscent of broccoli or cauliflower.

Availability

• Available all year, but most plentiful in fall and winter.

Selection

• Choose savoy salad with a good, fresh appearance.

Storage

• In a plastic bag, it can be kept for up to a week in the refrigerator, but it always tastes best when freshest.

Preparation

• Break off the leaves, discarding the heavy stalks at the base, then simply wash and dry, as you would any other salad green.

Suggestions

• Use several colourful heads of savoy salad as a centrepiece for your table and eat them the next day!

• Steamed or stir-fried with a little butter and salt, it makes a tasty side dish. (It won't lose colour unless overcooked.) Add a squeeze of lemon, if desired.

• Savoy salad is very attractive used to decorate or line a serving platter.

• To create a beautiful party dip, cut the base from a savoy salad and place it in a glass bowl, then set the bowl on a large tray or platter. Spread the central leaves out, putting a small bowl containing the dip right in the centre of the savoy salad. Arrange the vegetables or crackers around the bowl.

• Savoy salad is very pretty in a salad, alone or with other vegetables. It is excellent with Roquefort cheese dressing.

Nutritional Value
Savoy salad
(per 50 g, shredded = 1 3/4 oz)

			RDA**	%
calories*	(kcal)	12		
protein*	(g)	2.1		
carbohydrates*	(g)	3		
fat*	(g)	4		
fibre	(g)	0.4	30	1.3
vitamin A	(IU)	5200	3300	157
vitamin C	(mg)	69	60	115
thiamin	(mg)	0.02	0.8	2.5
riboflavin	(mg)	0.04	1.0	4
niacin	(mg)	0.4	14.4	2.8
sodium	(mg)		2500	
potassium	(mg)	190	5000	3.8
calcium	(mg)	92	900	10.2
phosphorus	(mg)	36.1	900	4
magnesium	(mg)		250	
iron	(mg)	1.6	14	11.4
water	(g)	46		

*RDA variable according to individual needs
**RDA recommended dietary allowance (average daily amount for a normal adult)

1
2
3
4
1. Boston
2. Red romaine
3. Green romaine
4. Green oak

Hydroponic lettuce

Origin

• All varieties of lettuce have a common ancestor that originated in Europe and Asia and are the most economically important members of the large Compositæ family. The word lettuce comes from the Latin word *lactuca,* which means milk, because the wild types of lettuce the Romans used secrete a milky substance when cut. This milky juice was used as a medicine for insomnia and nervous disorders and was considered to be a narcotic.

• The Assyrians and Egyptians cultivated the now rare asparagus lettuce, which has a phallic central stalk, hence their belief that it was an aphrodisiac. The Romans, on the other hand, grew head lettuce that they nicknamed "eunuch's salad". Nevertheless, salad with dressing became a fashionable way to end a meal during the reign of Emperor Augustus. When the Domitian became emperor, he moved the salad course to the beginning of the meal and people have been debating when to eat it ever since!

• Hydroponic lettuce has been developed only in the 1980s, with England leading the field in developing the technology. This lettuce is almost always grown with its roots in water instead of soil. There are about five major producer countries, including the U.S. and Canada.

Varieties and description

• Hydroponic lettuce is normally grown without pesticides, insecticides or herbicides and is extremely clean, as there is no soil in a hydroponic greenhouse. The main types of lettuce grown hydroponically are Boston and Bibb, which have soft, tender loosely packed leaves. These types are known as butterheads, to distinguish them from the crisper varieties, such as Iceberg.

• Long-leafed green and red hydroponic Romaine lettuce, often called baby Romaine because of its small size, is also popular. Small crops of red or green oak lettuce, lollo rossa — with curly red leaves showing a little green — and tango, another red curly-leafed variety, have recently been developed.

• A new hybrid called Novita will soon be out on the market. It's a cross between Boston and a curly leaf lettuce.

Availability

• Hydroponic lettuce can be found year round.

Selection

• Look for a nice green leaf that shows no signs of limpness or brown spots. Hydroponic lettuces are usually smaller and lighter than those grown in soil.

EXCELLENT SOURCE OF IRON
GOOD SOURCE OF VITAMINS A AND C
VERY LOW IN CALORIES, SODIUM AND FIBRE

Storage

• This lettuce keeps very well — up to two weeks when wrapped in plastic and refrigerated. However, all salad greens taste best when eaten fresh.

Preparation

• It is not necessary to wash hydroponic lettuce, unless you prefer to do so, as it is a very clean product. It's always best to tear a lettuce rather than cut it, for better absorption of dressing and to prevent discolouration of the cut ends.

• To prevent the lettuce from becoming soggy and limp, don't add salad dressing until you are ready to serve the salad.

Suggestions

• Use as a base for any salad and in sandwiches.

• Garnish soups with finely shredded lettuce.

• Blanch lettuce lightly in boiling water and serve as a delicate vegetable side dish, seasoned with melted butter and salt.

Nutritional Value

Boston lettuce

(per 100 g = 3½ oz)

			RDA**	%
calories*	(kcal)	13		
protein*	(g)	1		
carbohydrates*	(g)	2		
fat*	(g)			
fibre	(g)	0.3	30	1
vitamin A	(IU)	968	3300	29
vitamin C	(mg)	8	60	13
thiamin	(mg)	0.06	0.8	7.5
riboflavin	(mg)	0.06	1.0	6
niacin	(mg)	0.3	14.4	2.1
sodium	(mg)	9	2500	0.4
potassium	(mg)	264	5000	5.3
calcium	(mg)	35	900	3.9
phosphorus	(mg)	25	900	2.8
magnesium	(mg)	10	250	4
iron	(mg)	2	14	14.3
water	(g)	95		

*RDA variable according to individual needs
**RDA recommended dietary allowance (average daily amount for a normal adult)

8
7
6
5
4
3
2
1
1. Borage
2. Viola
3. Marigold
4. Sweet William
5. Pansy
6. Hyssop
7. Calendula
8. Carnation

Edible flowers

Origin

• Flowers and herbs have been used since ancient times for their unique flavour and appearance as well as for their medicinal properties. Certain flowers have been well known for centuries for their use in confectionary and liqueurs. Today, in California, edible flowers are becoming more popular as a gourmet treat. This influence is spreading throughout North America.

Varieties

• Among the many varieties of flowers being grown in North American greenhouses for culinary use are: Bergamot, Borage, Chive flowers, Chrysanthemums, Daisy flowers, Hibiscus, Honeysuckle, Impatiens, Jasmine, Lavender, Lime blossoms, Magnolias, Marigolds, Mimosa, Mustard flowers, Nasturtiums, Pansies, Petunias, Primroses, Red Poppies, Roses, Sweet Williams, Tagetes, Violas and Violets. You can expect to see additional varieties in the next few years.

Availability

• Some varieties are available year round and some have seasons. This is a very new industry and supplies may be scarce, as the restaurant trade uses most of the available harvest. Growers are currently increasing production, and we can expect to find them more easily in the future.

Selection

• There are many edible flowers in the world, but there are also many toxic ones. Flowers purchased from a florist have had chemicals used in all phases of their growth, and should NEVER be eaten. Nor should you eat flowers from your garden, unless you are an organic gardener and have done extensive research into which varieties are edible. Flowers grown for human consumption are chemical free and safe to eat. They are packaged in small plastic containers, and sold at specialty counters in fancier produce departments.

Storage

• You can keep your flowers for a few days in the refrigerator if necessary, but they are best used as fresh as possible.

Preparation

• These flowers are hydroponically grown and are very clean. Do not wash. Use as is, or separate the petals for particular uses, such as floral butters or crystalized petals. Remember to add the flowers right at the end of preparation, to preserve their appearance. If adding them to a salad, toss the dressing with the salad before you add the flowers, so that the petals won't have a chance to absorb the salad oil.

Suggestions

• Edible flowers can be used as a beautiful yet tasty garnish for meat, poultry or fish dishes, as well as some soups.

• Add to vegetable or fruit salads for a gourmet treat.

• Create a gorgeous, edible centrepiece for a special occasion.

• Float a few violet petals in champagne cocktails, or add red poppies to decorate a punch.

• Rose petals added to apple or peach pies give off a wonderful, appetizing aroma.

• Crystalized or used as is, flowers can lend a decorative touch to your desserts.

Edible flower salad

(serves 2-4)

Ingredients		
4	calendula flowers	
4	viola flowers	
4	marigolds	
125 g	mâche	1/4 lb
3	kumquats, sliced very finely in rounds	

Dressing ingredients		
30 mL	hazelnut or walnut oil	2 TBSP
30 mL	fresh orange juice	2 TBSP
15 mL	fresh lemon juice	1 TBSP
2.5 mL	honey	1/2 tsp
	pinch of salt and dash of pepper	

Instructions

- Arrange ingredients on individual plates or on a serving platter.
- Serve dressing separately, to preserve the beautiful appearance of the salad until the last moment.

Belgian endive

↑ Red endive

EXCELLENT SOURCE OF VITAMIN A
GOOD SOURCE OF VITAMIN C AND IRON
VERY LOW IN SODIUM
VERY LOW IN CALORIES

Origin

- Endives, sometimes called witloof (white leaf) or chicory, are members of the Compositæ family and closely related to escarole and chicory. They were discovered by accident in 1843 in Belgium, when some chicory roots germinated in a dark cellar. Today, this *blanching*, which gives endives their white colour and eliminates much of the bitterness, occurs when the plant is close to maturity. It is uprooted, drastically pruned, then transplanted and left to grow in darkness. They are not easy to cultivate, as everything must be done by hand through the many stages of growth. Endives were introduced into France in 1873, and to Holland a few decades later. These three countries are still the major producers, although Canada and the U.S. grow endives for local markets.

Varieties and description

- Endives are creamy white in colour, ending in pale yellow at the tip. They have a slightly bitter flavour and a crisp, refreshing texture. Red endives, a hybridization of radicchio rosso and white endive, have recently come onto the market. Red endives are slightly sweeter and should never be cooked or they will lose their distinctive colour and taste.

Availability

- Both red and white Belgian endives are available from September to May.

Selection

- Choose firm endives with tightly closed leaves and avoid those that are wilted or turning brown at the edges.

Storage

- Keep them refrigerated in a paper bag for no more than a few days. They are best when eaten fresh.

Preparation

- The grower peels away the outer leaves before the endive is sold for a nicer appearance, so there is no wastage for consumers. The cut edge may turn brown when exposed to air and left sitting, so do not prepare them before you are ready to use them. Endives are the cleanest vegetables on the market — washing should not be necessary. Discard a small slice from the end, then take apart leaf by leaf, slice in rounds or leave whole for cooking, depending on your recipe.

Suggestions

- Arrange the leaves decoratively, and serve with your favourite dip. For a party appetizer, put a dollop of sour cream in each leaf with 1 mL (1/4 teaspoon) of caviar.
- Add endives to a salad.
- Try cream of endive soup.
- Endives are delicious braised, baked or steamed.

↓ White endive

Endive and mandarin salad

(serves 4)

Ingredients		
4	endives	
2	mandarins or tangerines, peeled and separated into segments	
125 mL	walnuts, coarsely chopped	1/2 cup
1	small head of lettuce, torn into bite-sized pieces	
	Dressing	
	juice of 1 mandarin	
15 mL	fresh lemon juice	1 TBSP
125 mL	plain yogurt	1/2 cup
	salt and white pepper to taste	

Instructions

- Separate a dozen whole endive leaves and slice the rest.
- Mix salad ingredients together with dressing, and decorate with whole endive leaves.

Nutritional Value
Endive
(per 100 g = 3 1/2 oz)

			RDA**	%
calories*	(Kcal)	20		
protein*	(g)	1.7		
carbohydrates*	(g)	4.1		
fat*	(g)	0.1		
fibre	(g)	0.9	30	3
vitamin A	(UI)	3300	3300	100
vitamin C	(mg)	10	60	17
thiamin	(mg)	0.07	0.8	9
riboflavin	(mg)	0.14	1.0	14
niacin	(mg)	0.5	14.4	4
sodium	(mg)	14	2500	0.6
potassium	(mg)	294	5000	5
calcium	(mg)	81	900	9
phosphorus	(mg)	54	900	6
magnesium	(mg)	10	250	4
iron	(mg)	1.7	14	12
water	(g)	93		

*RDA variable according to individual needs
**RDA : recommended dietary allowances (average daily amount for normal adult)

Jicama

(pronounced hee-ca-ma)

GOOD SOURCE OF POTASSIUM
GOOD SOURCE OF VITAMINS B AND C
LOW IN CALORIES AND SODIUM

Origin

• Jicama, also known as Mexican turnip, ahipa and yam bean, is a tuber from a leguminous plant of the Pulse family, that originated in Mexico as well as the Amazon region of South America. Its seeds were used medicinally by the Aztecs, who named it xicamatl. Spanish explorers introduced jicama into the Philippines in the 17th century, and it gradually spread through Pacific countries to Asia, particularly China. Mexico is one of the main exporters of jicama today.

Varieties and description

• Jicama looks like a turnip that has been slightly flattened on both ends, with a thin light brown skin and white flesh. It can weigh from 250 g to more than 2.25 kg (1/2 to 5 lbs) It is crisp and crunchy, very juicy, and slightly sweet — reminiscent of water chestnuts.

Availability

• Jicama is available year round, though it is more plentiful from December to April.

Selection

• Medium to small jicama are juicier and less fibrous. Select clean-looking, firm jicama, avoiding cuts and dark spots.

Storage

• Store jicama as you would store potatoes — in a cool dry place — and it will stay fresh for several weeks. You can cut part of it to use, then store the rest (wrapped in plastic) in the refrigerator for up to a week.

Preparation

• Simply peel the inedible skin, and it is ready to use. It is delicious raw or cooked. To preserve its crispy texture, cook it very lightly.

Suggestions

• Try jicama Mexican-style, sliced thinly in rounds, with lemon or lime juice squeezed over it, sprinkled with chili powder and salt. A tasty snack!

• Sticks of jicama are great with your favourite dip.

• Dice jicama into any salad, including fruit salads.

• Jicama is an excellent substitute for water chestnuts when used in a stir-fry.

• Jicama is wonderful in sweet and sour dishes.

Nutritional Value

Jicama

(per 100 g = 3 1/2 oz)

			RDA**	%
calories*	(Kcal)	22		
protein*	(g)	0.9		
carbohydrates*	(g)	11.34		
fat*	(g)	0.03		
fibre	(g)	0.47	30	1.5
vitamin C	(mg)	12.7	60	21
thiamin	(mg)	0.08	0.8	10
riboflavin	(mg)	0.07	1.0	7
niacin	(mg)	0.3	14.4	2
sodium	(mg)	7	2500	0.3
potassium	(mg)	250	5000	5
calcium	(mg)	9	900	1
phosphorus	(mg)	20	900	2
magnesium	(mg)	18	250	7
iron	(mg)	1.9	14	7
water	(g)	93		

*RDA variable according to individual needs
**RDA : recommended dietary allowances (average daily amount for normal adult)

Jicama and mango salad

(serves 4)

Ingredients		
1	small or medium jicama	
2	small mangos, peeled and diced	
1	romaine lettuce	
30-45 mL	sesame seeds, toasted	2-3 TBSP

Instructions

- Peel and chop the jicama into thin french-fry shapes.
- Wash the romaine lettuce, dry and cut in strips.
- In a salad bowl or on a serving platter, arrange the jicama and mangos on top of the lettuce.
- Add the dressing, mix well, garnish with sesame seeds and serve immediately.

Dressing ingredients		
45 mL	sunflower seed oil	3 TBSP
30 mL	fresh lime juice	2 TBSP
30 mL	fresh orange juice	2 TBSP
5-10 mL	honey	1-2 tsp
	salt and white pepper to taste	

Instructions

- Mix well.

Fennel

GOOD SOURCE OF VITAMINS A AND C
VERY LOW IN CALORIES
LOW IN FIBRE
TRACES OF CALCIUM AND IRON

Origin

• Sweet or Florence fennel, also known by its Italian name, finocchio, is popular throughout the Mediterranean. The ancient Greeks prized it as a medicine and as a diet food and the Romans cultivated it extensively, more for the fine-leafed tops, which they often used as a flavourful herb, than for the bulbous vegetable part. They also made an extract of fennel, which they used in treating eye diseases, especially cataracts. Italian immigrants introduced the plant to the United States, where it is now grown for export. Fennel is sometimes mistakenly called anise.

Varieties and description

• Fennel is a perennial of the Umbelliferæ family, which includes parsley, carrots and celery. It does, in fact, look like a bulbous truncated celery. The slightly sweet flavour (anethole) will remind you of licorice or anise, but is pleasanter and much fainter. The bulb is usually about the size of a grapefruit and ranges in colour from pale green to white.

Availability

• Available all year, the main season for fennel is from October to April.

Selection

• Choose firm fennel with a good appearance. It's an indication of freshness if the tops are in good condition.

Storage

• Fennel should be used within a few of days of purchase or the flavour will begin to fade and it will start to dry out. Separate the tops, which spoil at a faster rate, keeping the stalks to use in salads or cooked dishes. Wrap the bulbs in plastic and refrigerate. If your fennel becomes limp, revive it in a bowl of very cold water.

Preparation

• Discard a thin slice from the base and trim off the stalks to about 3 centimetres (1 inch) from the bulb. Wash well. Use a vegetable peeler on the outer stalks only if necessary. Depending on your recipe, cut thin rounds or chunks (for salads), halve (for cooking), slice on the diagonal (for stir-fry), julienne, mince, etc.

Suggestions

• Fennel's crunchy texture makes it a great addition to a salad.

• Braised, baked, creamed, grilled or steamed, with a little butter or parmesan and salt, fennel is an excellent side dish.

• Stir-fry it with other Mediterranean vegetables.

• It's delicious dipped in batter and fried, tempura-style.

• Fennel is excellent in soups.

• Fennel gives a wonderful flavour to fish and seafood dishes.

Nutritional Value

Fennel

(per 100 g = 3½ oz)

			RDA**	%
calories*	(Kcal)	28		
protein*	(g)	2.8		
carbohydrates*	(g)	5.1		
fat*	(g)	0.4		
fibre	(g)	0.5	30	2
vitamin A	(UI)	3500	3300	106
vitamin C	(mg)	31	60	52
potassium	(mg)	397	5000	8
calcium	(mg)	100	900	11
phosphorus	(mg)	51	900	6
iron	(mg)	2.7	14	19

*RDA variable according to individual needs
**RDA : recommended dietary allowances (average daily amount for normal adult)

Fennel and tomato salad

(serves 4)

Ingredients		
2	fennel, sliced in thin rounds	
1	medium sweet onion, sliced in thin rounds	
2	tomatoes, cut in quarters	
8	whole lettuce leaves	
	Dressing	
30 mL	olive oil	2 TBSP
30 mL	white wine vinegar	2 TBSP
1	garlic clove, finely minced	
5 mL	fresh mint, finely chopped	1 tsp
	salt and freshly ground black pepper to taste	

Instructions

- Mix the dressing well and toss together with fennel, onion and tomato slices. Serve on a bed of whole lettuce leaves.

Fennel salad Waldorf

(serves 4)

Ingredients		
2	medium fennel bulbs, sliced thinly	
1	red apple, diced	
125 mL	walnuts, chopped coarsely	1/2 cup
4	whole lettuce leaves, preferably Boston lettuce	
	Dressing	
45 mL	olive oil	3 TBSP
30 mL	balsamic or raspberry vinegar	2 TBSP
15 mL	mayonnaise	1 TBSP
	salt and pepper to taste	

Instructions

- Mix dressing ingredients together well. Toss salad and dressing. Serve each portion on top of a whole lettuce leaf.

↓ Fennel and tomato salad

Radicchio

(pronounced ra-dee-key-o)

EXCELLENT SOURCE OF VITAMIN A
CONTAINS VITAMIN C
SOURCE OF POTASSIUM
LOW IN CALORIES

Origin

• Radicchio, often called red chicory, is a member of the large Compositæ family. It is originally from the province of Veneto in northern Italy, where it once grew wild. The Italians have been cultivating it since the 16th century and are still the major source of this beautiful red head lettuce, although other countries in Central and South America are now starting to grow it for export. It is the only member of the chicory salad family that turns red when cool weather prevails.

Varieties and description

• The best known variety is dark ruby-coloured, with rounded, shiny, smooth, white ribbed leaves. Shaped like a small cabbage, though the leaves are much looser, it is more like a full-blown rose. Other varieties vary in shapes of leaves, and may contain flecks of pink, red or green, but the taste remains the same — slightly sharp and agreeably bitter, like white endive or escarole.

Availability

• Italian radicchio is harvested from November to mid-February.

Selection

• Look for a compact, firm head, with no brown discolouration at the tip of the leaves or at the base.

Storage

• Keep in the refrigerator, in a plastic bag. It tastes best when crisp and fresh but can be kept for up to a week.

Preparation

• There should be nothing to discard, other than the small core, if necessary. Simply wash and dry.

Suggestions

• This highly decorative leaf can be used to garnish many dishes.

• A single leaf can hold a serving of an appetizer, such as shrimp salad or celeriac salad.

• Serve it Italian-style with olive oil, wine vinegar, salt and pepper. Adding even a leaf or two chopped in small pieces will turn an ordinary green salad into a gourmet treat.

• Radicchio is sometimes cooked in Italian cuisine; in pasta dishes, in omelets, sautéed lightly, or even breaded and fried.

Nutritional Value

Radicchio

(per 100 g = 3½ oz)

			DRA**	%
calories*	(Kcal)	20		
protein*	(g)	1.8		
carbohydrates*	(g)	3.8		
fat*	(g)	0.3		
fibre	(g)	0.8	30	3
vitamin A	(UI)	4000	3300	121
vitamin C	(mg)	22	60	37
thiamin	(mg)	0.06	0.8	8
riboflavin	(mg)	0.1	1.0	10
niacin	(mg)	0.5	14.4	4
potassium	(mg)	420	5000	8
calcium	(mg)	86	900	10
phosphorus	(mg)	40	900	4
magnesium	(mg)	13	250	5
iron	(mg)	0.9	14	6
water	(g)	94		

*RDA variable according to individual needs
**RDA : recommended dietary allowances (average daily amount for normal adult)

Radicchio and smoked salmon hors d'oeuvre

(serves 4)

Ingredients	
1	head of radicchio, torn in bite-sized pieces
1	head of Boston lettuce, torn in bite-sized pieces
1	grapefruit or 2 oranges, peeled
	orange and lime zest, to garnish
8	green olives, sliced
4	slices of smoked salmon to garnish

Dressing ingredients		
45 mL	olive oil	3 TBSP
30 mL	lemon juice	2 TBSP
10 mL	fresh chives, finely minced	2 tsp
	salt and freshly ground black pepper to taste	

Instructions

- Segment the citrus and remove all the membrane.
- Mix dressing well.
- Toss all ingredients except smoked salmon with the dressing in a salad bowl.
- Garnish with the slices of smoked salmon, or arrange on four plates ready to serve.

Hydroponic tomatoes

GOOD SOURCE OF VITAMINS A AND C
LOW IN CALORIES
VERY LOW IN SODIUM AND FIBRE

Origin

• The tomato, originally a tiny wild fruit from Peru, is a member of the Solanaceæ family. Tomatoes gradually spread through pre-Columbian cultures before the arrival of the conquistadors. Returning explorers brought the larger cultivated varieties to Europe, where they were regarded with great suspicion for about two centuries. Called love apples, they were thought to excite violent passions. Sir Walter Raleigh started a fad when he gave Elizabeth I some tomato plants and people began to grow them as an ornamental curiosity. Hydroponic tomatoes have been developed in the past decade or two in several countries with colder climates, as an alternative to normal greenhouse cultivation. These tomatoes are usually grown in nutrient-fed water from start to finish, with air pumped through the water to provide oxygen to the roots. These plants must be supported to keep the plants upright. Some growers use other substitutes for soil, such as peat moss, vermiculite, rock wool, sawdust, etc. The tomatoes are then fed proper nutrients, sometimes several times a day. This controlled environment helps prevent the tomatoes from being attacked by insects, and eliminates the need for pesticides.

Varieties and description

• These juicy, fleshy tomatoes are perfect looking and uniform in size. They are ripe and ready to eat when picked.

Availability

• Available most of the year.

Selection

• Hydroponic tomatoes are wrapped and packaged in single-layer trays and shipped with utmost care, so there is never a problem to select them. Please note that the stems are very hard, and could puncture the tomato next to them if the tomatoes are stacked on top of each other, or if not carefully placed in a bag.

Storage

• You can keep these tomatoes for five to seven days at room temperature. It is better not to refrigerate them because the cold ruins their vine-ripened flavour.

Preparation

• As with any other tomato, simply wash and slice. We do not recommend cooking hydroponic tomatoes, not because they wouldn't be delicious, but because they are so tasty, they deserve to be eaten in their natural state.

Suggestions

• Try a vinaigrette containing finely minced dry shallots over sliced hydroponic tomatoes. Garnish with fresh dill.

• Prepare an Italian salad; slice tomatoes and mozzarella, add some fresh basil (finely chopped), drizzle olive oil and a little lemon juice over it, and add a small amount of freshly ground black pepper and salt.

• Try stuffing tomatoes with tuna or shrimp salad.

• Slice your tomatoes, then mash a few anchovies into 45 or 60 mL (3 or 4 tablespoons) of sour cream, and put a dollop on each slice. Garnish with olives.

Tomato, avocado and shallot salad

(serves 4)

Ingredients	
4	tomatoes, sliced
2	avocados, peeled and cut in quarters
6 to 8	dry shallots, peeled and cut in paper-thin rings

Dressing ingredients		
45 mL	olive oil	3 TBSP
30-45 mL	lemon juice or white wine vinegar	2-3 TBSP
1	garlic clove, finely minced	
15 mL	fresh thyme, finely chopped	1 TBSP
salt and freshly ground black pepper to taste		

Instructions

- Prepare dressing ahead of time to allow the flavour of the thyme to permeate the oil.
- Arrange tomatoes, avocados and shallots on a serving platter.
- Pour dressing over the vegetables, and garnish with sprigs of fresh thyme.

Nutritional Value
Hydroponic tomatoes
(per 100 g = $3^1/_2$ oz)

			RDA**	%
calories*	(Kcal)	22		
protein*	(g)	1.1		
carbohydrates*	(g)	5		
fat*	(g)	0.2		
fibre	(g)	0.5	30	2
vitamin A	(UI)	900	3300	27
vitamin C	(mg)	23	60	38
thiamin	(mg)	0.06	0.8	7.5
riboflavin	(mg)	0.04	1.0	5
niacin	(mg)	0.7	14.4	4
sodium	(mg)	3	2500	0.1
potassium	(mg)	244	5000	4
calcium	(mg)	13	900	1.4
phosphorus	(mg)	27	900	3
magnesium	(mg)	14	250	6
iron	(mg)	0.5	14	4
water	(g)	93.5		

*RDA variable according to individual needs
**RDA : recommended dietary allowances (average daily amount for normal adult)

Mâche

(pronounced mash)

Low in calories
Source of magnesium
Low in fibre

Origin

• Mâche is also known as lamb's lettuce, field salad or corn salad. A member of the Compositæ family, which includes dandelions, chicory, endives and artichokes, mâche is very closely related to other types of lettuce. It is a hardy, weed-like annual of Mediterranean origin, regarded by Europeans as a delicacy since early Roman times. Today, France and Holland are major producers of mâche. Recently, Canada and the U.S. have increased their production.

Varieties and description

• The sweet, nutty-tasting, tongue-shaped leaves are in small clusters and range from a medium to dark green. The long-leafed variety is called "blond" or "green" mâche, and the short-leafed one, known as "shell" mâche, is darker, firmer, and stronger tasting and therefore more popular.

Availability

• Available year round where grown hydroponically or in ordinary greenhouses, its natural season is from winter to spring.

Selection

• Mâche is usually picked in small clusters with the roots still attached, and sold in small plastic trays. It should look fresh and green.

Storage

• It should be eaten as fresh as possible, or it will spoil. Keep mâche wrapped in the refrigerator and use within a few days.

Preparation

• Mâche needs to be well washed because it is usually grown in sandy soil, but handle it very gently and dry with care. If you wash it ahead of time, refrigerate until used. Cut off the tiny roots before washing. Do not add dressing until the moment you are ready to eat, or it will get soggy and lose its fresh flavour.

Suggestions

• Use it in a salad on its own or with another soft, mild lettuce such as Boston or Bibb. Try an oil such as walnut, pumpkin seed, or hazelnut to compliment the nutty flavour of mâche, with a squeeze of lemon juice.

• Cut mâche into bite-sized pieces and use to garnish a cream soup.

• If you like British tea sandwiches — very thin, fresh sandwich bread, lightly spread with butter or a soft creamy cheese, with fillings such as paper-thin cucumber or watercress — try them with mâche for a delicious treat.

Nutritional Value

Mâche

(per 100 g = 3 1/2 oz)

			RDA**	%
calories*	(Kcal)	21		
protein*	(g)	2		
carbohydrates*	(g)	3.6		
fat*	(g)	0.4		
fibre	(g)	0.8	30	3
magnesium	(mg)	13	250	5
water	(g)	93		

*RDA variable according to individual needs
**RDA : recommended dietary allowances (average daily amount for normal adult)

Fancy mâche salad

(serves 4)

Ingredients		
3 to 4	endives	
125 g	mâche, left whole	1/4 lb
125 mL	pecans, chopped coarsely	1/2 cup
12 to 16	edible Sweet Williams	

Dressing ingredients		
45 mL	walnut oil	3 TBSP
15 mL	fresh lemon juice	1 TBSP
15 mL	fresh orange juice	1 TBSP
5 mL	lemon zest	1 tsp
1-2 mL 5 mL	anise seeds or anise-flavoured liqueur	1/4 tsp 1 tsp
	salt and pepper to taste	

Instructions

- Keep a few endive leaves whole for decoration and slice the rest.
- Arrange the endives, mâche and nuts on a serving platter.
- Mix the dressing ingredients well.
- Gently mix in dressing when ready to serve.
- Add flowers as a final decorative touch.

Sweet coloured peppers

EXCELLENT SOURCE OF VITAMINS A AND C
HIGH IN POTASSIUM
GOOD SOURCE OF FIBRE
LOW IN CALORIES AND SODIUM

Origin

• Capsicums, or sweet peppers, members of the Solanaceæ family, are originally from the tropical areas of North, Central and South America. They were brought back to Europe by Columbus and other explorers, and later they became very popular, especially in Mediterranean cooking. In recent years, the Dutch, using hydroponic growing methods, have created new hybrids of exotic coloured peppers. The Dutch are the foremost experts on this appealing product. Some of the exotic colours are also grown in greenhouses in Canada and the U.S., and some are grown in the field in Mexico.

Varieties and description

• In addition to the green and red peppers that are fairly well known, you can now find white (sometimes called Hungarian peppers), yellow, orange, purple, violet, brown and black peppers. The Dutch are working on two new varieties: pink and lilac! Green peppers are actually harvested before they have reached their peak of ripeness, sweetness and full nutritional value. All coloured peppers start off as green peppers. Any normal green pepper will turn red if left to ripen on the plant. Purple, brown and black peppers, which also start out green, will revert back to green if left too long on the plant after they have ripened. (They should not be cooked, as their colour would change back to green.)

• As for taste, purple peppers taste similar to mild green peppers, white ones are rather like yellow ones, and brown ones are exceptionally sweet. Generally, orange and red peppers are the sweetest.

Availability

• Generally available all year round, especially green, red and yellow peppers. The other colours may be scarce at times, but are becoming less so as more countries begin production of exotic colours. The Dutch peppers are available from April to December.

Selection

• Choose a good-looking pepper with a firm, unwrinkled skin.

Storage

• Peppers keep well refrigerated for about a week.

Preparation

• Slice in half and remove the stalk and seeds. If you are going to stuff it, cut carefully around the stalk and pull it out, clean out the seeds, stuff it and replace the top like a removable cover. (If it won't stand up, take a thin slice off the bottom.)

Suggestions

• Sliced in strips or rounds, in any salad or with a dip, coloured peppers add a crisp, flavourful, decorative touch.

• Use in stir-fries, or in grilled or barbecued kebabs.

• Stuff peppers with rice and meat filling and bake, or stuff with salad mixture (salmon salad, cottage cheese, etc.) and eat raw.

• Use coloured peppers in a great variety of Mediterranean dishes, such as ratatouille.

• Coloured peppers look great on pizza.

• Roast peppers under the broiler, then place them in a paper bag for 5 minutes (the steam that is created will loosen the skins). Skin, remove seeds and slice, then marinate in vinaigrette.

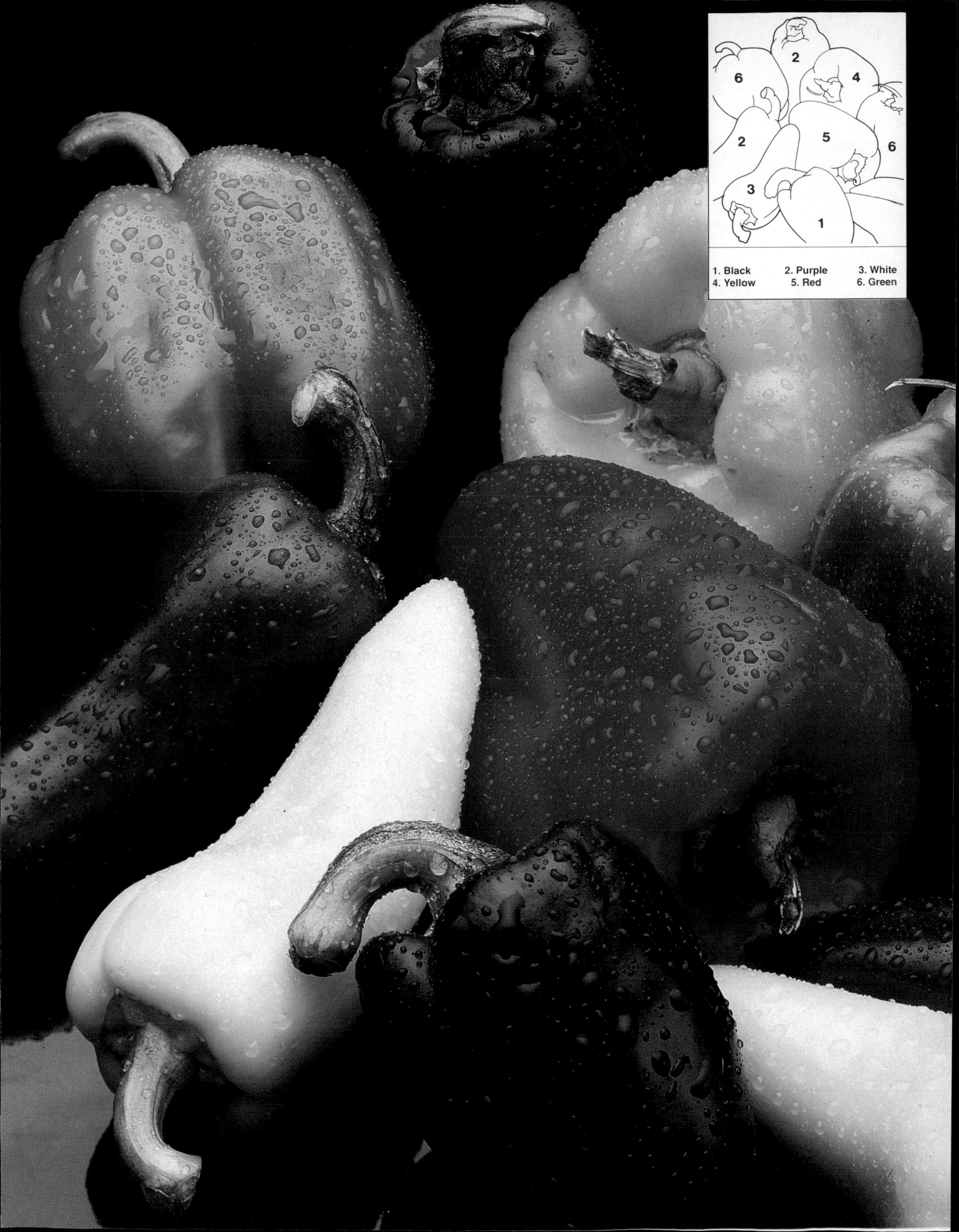

2
6
4
2
5
6
3
1
1. Black
2. Purple
3. White
4. Yellow
5. Red
6. Green

Pepper pasta salad

(serves 4)

Ingredients		
250 g	small pasta shapes (shells, twists, etc.)	8 oz
15 mL	olive oil	1 TBSP
3	different coloured peppers, diced	
125 mL	olives, sliced	1/2 cup
60 mL	scallions or sweet onion, finely chopped	1/4 cup
	small handfull of parmesan cheese*	
1	whole basil leaf to garnish	

Dressing ingredients		
60 mL	olive oil	4 TBSP
45 mL	white wine vinegar	3 TBSP
1	garlic clove, finely minced	
45 mL	parsley or chervil, finely chopped	3 TBSP
15 mL	fresh basil, finely chopped	1 TBSP

** Or 1 small, fresh, soft goat cheese (covered with black pepper, or fine herbs), crumbled into small chunks.*

Instructions

- Cook the pasta, rinse in cold water and drain well. Mix a spoon of olive oil into the pasta to prevent it from sticking together and let it cool.
- Dice the peppers, using as many different coloured peppers as possible. Use only a half or third of each, depending on how many colours you can find.
- Mix the dressing ingredients well.
- Toss dressing gently into salad ingredients. Serve as is or on a bed of alternating radicchio and green lettuce leaves.

Nutritional Value
Sweet coloured peppers
(per 100 g = 3 1/2 oz)

			RDA**	%
calories*	(Kcal)	24		
protein*	(g)	1		
carbohydrates*	(g)	4		
fat*	(g)	trace		
fibre	(g)	1.4	30	5
vitamin A	(UI)	419	3300	13
vitamin C	(mg)	127	60	212
thiamin	(mg)	0.08	0.8	10
riboflavin	(mg)	0.08	1.0	8
niacin	(mg)	0.5	14.4	3.5
sodium	(mg)	12	2500	0.5
potassium	(mg)	212	5000	4
calcium	(mg)	9	900	1
phosphorus	(mg)	22	900	2.4
iron	(mg)	0.6	14	4
water	(g)	93.4		

*RDA variable according to individual needs
**RDA : recommended dietary allowances (average daily amount for normal adult)

Red peppers have 10 times more vitamin A and twice as much vitamin C than green peppers, yellow ones have even more than red, and orange ones have the most.

Marinated pepper salad

(serves 4)

Ingredients
3 sweet peppers of different colours, cut in strips
whole lettuce leaves

Dressing ingredients		
45 mL	olive oil	3 TBSP
45 mL	lemon juice or white wine vinegar	3 TBSP
2.5 mL	Dijon mustard	1/2 tsp
	salt and pepper to taste	

Instructions

- Boil the strips of pepper about 3 minutes, pour cold water over them and drain well. Let cool.
- Mix the dressing, add to the peppers, and let stand at room temperature to marinate for at least an hour.
- Serve on a bed of whole lettuce leaves.

Sprouts

SOURCE OF PROTEIN
SOURCE OF VITAMIN B COMPLEXE
VERY LOW IN CALORIES, SODIUM AND FIBRE

1. Radish sprouts 2. Mustard sprouts

Origin

• Mung and soybean sprouts have been popular vegetables in Oriental countries for many centuries, probably because they are so fast growing and easy to produce. What other fresh vegetable can be grown in the middle of winter, is a good source of fibre and nutrition, and doesn't even need a greenhouse or field to grow in?

• Alfalfa (also called lucerne) and the peppery-tasting sprouting cresses have long been popular in western countries for the same reasons.

Varieties and description

• Sprouts are the tender young shoots from seeds that germinate very quickly. Only a few kinds of sprouts are available commercially, such as the more familiar mung bean and the slightly stronger-tasting soybean sprouts (legumes), which are usually cooked, or the tiny alfalfa sprouts, mustard cress, radish sprouts and garden cress, which are usually eaten raw.

Availability

• Available year round.

Selection

• Choose sprouts that are fresh and crispy, and show no signs of discolouration.

Storage

• Refrigerate in a plastic bag and use within a day or two of purchase.

Preparation

• Simply wash and drain, or pat dry with a paper towel. If they seem a bit limp, revive them by soaking in very cold water for a few minutes. All sprouts taste good raw, but not all sprouts can be cooked.

Suggestions

• Sprouts taste crispy and refreshing in a salad.

• Use them as an edible garnish.

• Mung bean and soybean sprouts are excellent in stir-fries.

• Alfalfa sprouts, radish sprouts or sprouted cress are tasty in most sandwiches. Try them in a pita pocket sandwich.

Nutritional Value
Mung bean sprouts
(per 100 g = 3½ oz)

			RDA**	%
calories*	(Kcal)	35		
protein*	(g)	3.8		
carbohydrates*	(g)	6.6		
fat*	(g)	0.2		
fibre	(g)	0.7	30	2
vitamin A	(UI)	20	3300	0.6
vitamin C	(mg)	19	60	32
thiamin	(mg)	0.13	0.8	16
riboflavin	(mg)	0.13	1.0	13
niacin	(mg)	0.8	14.4	6
sodium	(mg)	5	2500	0.2
potassium	(mg)	223	5000	4
calcium	(mg)	19	900	2
phosphorus	(mg)	64	900	7
iron	(mg)	1.3	14	9
water	(g)	91		

*RDA variable according to individual needs
**RDA : recommended dietary allowances (average daily amount for normal adult)

Oriental soybean sprout salad

(serves 4)

Ingredients		
250 g	fresh sprouts	8 oz
1	red pepper, cut in small slivers	
1	carrot, grated	
	a few snow peas (blanched or uncooked)	
	handful of toasted (unsalted) cashew nuts	
	Dressing	
60 mL	sesame seed oil	4 TBSP
30 mL	rice vinegar	2 TBSP
15 mL	soya sauce	1 TBSP
7 mL	fresh ginger, minced	1/2 TBSP
1	garlic clove, minced	
2 mL	sugar	1/2 tsp
	pepper to taste	
	small handful of finely chopped scallions	

Instructions

- Blanch the sprouts for a minute or two in boiling water.
- Cut the snow peas into thin slivers on the diagonal.
- Toss dressing in salad and let marinate about half an hour, mixing several times.
- Add nuts to garnish just before serving.

Spinach and radish sprout salad

(serves 4)

Ingredients		
1	package of spinach leaves	
1	bunch of radish sprouts	
1	English cucumber, sliced finely	
125 mL	almonds toasted with a sprinkling of tamari or soy sauce until browned	1/2 cup
	Dressing	
60 mL	cottage cheese	4 TBSP
30 mL	fresh lemon juice	2 TBSP
15 mL	olive oil	1 TBSP
	salt and pepper to taste	

Instructions

- Wash the spinach, then dry and remove the stems.
- Prepare sprouts, cut off just above the roots.
- Mix dressing and toss together with ingredients in a large salad bowl.
- You may want to garnish with slivers of yellow pepper or carrot for some added colour.

Spinach and radish sprout salad →

Watercress

Origin

• A member of the Crucifer family of plants, watercress originated in the Mediterranean area and in Asia Minor and was later brought by early settlers to North America, where it now grows in most areas. Watercress was once thought to be a cure for madness. The Greeks thought it helped to improve one's mental ability.

Varieties and description

• Watercress is a creeping plant, whose roots thrive in cold, clear running water or in prepared beds. It has long stems that branch out into clusters of small dark green, rounded leaves and a pleasantly pungent zesty taste.

Availability

• It can be found all year round, but is most plentiful in the spring and summer.

Selection

• As with other salad greens, look for crisp, fresh green-leafed cress. Watercress is usually sold tied in bunches standing up in water.

Storage

• Undo the bunch and wash the cress in very cold water.

EXCELLENT SOURCE OF VITAMINS A AND C
SOURCE OF CALCIUM AND IRON
VERY LOW IN CALORIES

Dry well between paper towels or by spinning, and put in a plastic bag with a piece or two of paper towelling before refrigerating. It will keep for a few days, but is best when eaten fresh.

Preparation

• Discard a small piece from the base of the stems. Chop into bite-sized pieces for salads or sandwiches. Leave whole for garnishes.

Suggestions

• Watercress is a delightful addition to any salad.

• Watercress soup is excellent. Make a simple version of leek and potato soup, then add a bunch of coarsely chopped watercress before the soup is puréed. Garnish each bowl with finely chopped cress. Watercress is a great garnish for any puréed soup.

• Mince it finely and add to a sour cream or soft cheese dip for raw vegetables or crackers.

• Add watercress to a cheese or tomato sandwich.

• Watercress is very good in a stir-fry. (Add at the last minute to avoid over-cooking.)

• Watercress is a very attractive garnish.

Nutritional Value

Watercress
(per 100 g = 3½ oz)

			RDA**	%
calories*	(Kcal)	19		
protein*	(g)	2.2		
carbohydrates*	(g)	3.0		
fat*	(g)	0.3		
fibre	(g)	0.7	30	2.3
vitamin A	(UI)	4900	3300	148
vitamin C	(mg)	79	60	131
thiamin	(mg)	0.08	0.8	10
riboflavin	(mg)	0.16	1.0	16
niacin	(mg)	0.9	14.4	6
sodium	(mg)	52	2500	2
potassium	(mg)	282	5000	6
calcium	(mg)	151	900	17
phosphorus	(mg)	54	900	6
magnesium	(mg)	20	250	8
iron	(mg)	1.7	14	12
water	(g)	93		

*RDA variable according to individual needs
**RDA : recommended dietary allowances (average daily amount for normal adult)

Watercress and avocado salad

(serves 4-8)

Ingredients		
1	bunch watercress, torn in pieces	
2	ripe avocados	
	lemon juice	
3	endives, coarsely chopped	
250 g	mushrooms, sliced finely	1/2 lb
	edible marigolds (optional)	
	Dressing	
60 mL	sunflower or olive oil	4 TBSP
45 mL	fresh lemon juice	3 TBSP
5 mL	Dijon mustard	1 tsp
1	garlic clove, finely minced	
	salt and freshly ground black pepper to taste	

Instructions

• Peel the avocados, cut in quarters and sprinkle with lemon juice.

• Mix dressing well.

• Mix all ingredients together, except flowers, in a large bowl or on a serving platter and add dressing just before serving.

• Add flowers at the last minute, to keep the petals from absorbing salad oil.

↑ Watercress and avocado salad

Watercress dressing

(for the salad of your choice)

Ingredients		
250 mL	watercress leaves, finely minced	1 cup
500 mL	natural yogurt	2 cups
	small handful of scallions (green onions), finely minced	
1	garlic clove, crushed	
	salt and pepper to taste	
	squeeze of lemon	

Instructions

• Mix all ingredients well, and use as a dressing for avocados, sweet coloured peppers, cucumbers and tomatoes, or the salad of your choice. Keep any leftover dressing in the refrigerator for your next salad.

Arugula

(pronounced a-roo-goo-la)

EXCEPTIONAL SOURCE OF VITAMIN C
EXCELLENT SOURCE OF IRON AND CALCIUM
LOW IN CALORIES

Origin

• Also known as salad rocket, roquette, rugula, rucola and Mediterranean rocket, arugula, of the Cruciferæ family, has long been popular with Italians. It originated in the Mediterranean area and was introduced to North America by Italian immigrants. Today, arugula is cultivated in greenhouses in Canada and the U.S. as well as in Europe.

Varieties and description

• Closely related to the radish, these fine, dark green, smooth, irregularly shaped leaves have a pleasantly sharp taste when young and fresh, reminiscent of watercress, with a nutty, peppery taste.

Availability

• Available all year round.

Selection

• Only tender young leaves are recommended, as the older ones are tough and very bitter. The leaves should be bright and fresh looking, with no signs of yellowing, limpness or dark spots.

Storage

• It should be used as fresh as possible. If you have to keep it for a day or two, dampen it, wrap it in a paper towel, and put it in a plastic bag before refrigeration.

Preparation

• Cut off the little rootlets and any stems that seem coarse. Do not wash until just before using, then dry well. Arugula is best used along with other salad vegetables, as its tart flavour combines well with milder types of lettuce.

Suggestions

• Use arugula in salads.

• Add arugula to puréed soups for a special flavour.

• Arugula is an attractive garnish.

• Arugula is very tasty in sandwiches.

Nutritional Value

Arugula

(per 100 g = 3 1/2 oz)

			RDA**	%
calories*	(Kcal)	—		
protein*	(g)	0.6		
carbohydrates*	(g)	4.1		
fat*	(g)	0.1		
fibre	(g)	0.8	30	2.6
vitamin A	(UI)	1225	3300	37
vitamin C	(mg)	120	60	200
thiamin	(mg)	0.2	0.8	25
riboflavin	(mg)	0.3	1.0	30
niacin	(mg)	1.2	14.4	8.3
calcium	(mg)	205	900	22.8
phosphorus	(mg)	3.7	900	0.4
iron	(mg)	9.5	14	67.9

*RDA variable according to individual needs
**RDA : recommended dietary allowances (average daily amount for normal adult)

Arugula and radicchio salad

(serves 4)

Ingredients		
1 or 2	bunches of arugula, left whole	
1	head of radicchio, torn in bite-sized pieces	
	about half a head of chicory lettuce torn in bite-sized pieces	
1	orange, peeled and segmented or cut in slices	
	violas to garnish (optional)	
Dressing		
30 mL	olive oil	2 TBSP
30 mL	balsamic or red wine vinegar	2 TBSP
	salt and freshly ground black pepper to taste	

Instructions

- Arrange salad ingredients on a serving platter or in a bowl.
- Mix dressing well, then toss together with salad.
- Add viola flowers and serve immediately.

Arugula Italian-style salad

(serves 4)

Ingredients		
1 or 2	bunches of arugula, torn in bite-sized pieces	
2	tomatoes, sliced in thin rounds	
1	Boston lettuce, torn in bite-sized pieces	
1	handful scallions, finely chopped	
1/4	sweet onion, finely sliced	
Dressing		
	juice of 1 lemon	
60 mL	olive oil	4 TBSP
1	garlic clove, finely minced	
15 mL	fresh oregano, finely chopped	1 TBSP
	freshly ground black pepper and salt to taste	

Instructions

- Mix all ingredients and pour dressing over vegetables.

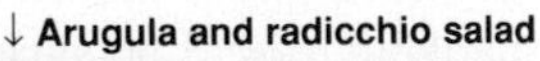

↓ Arugula and radicchio salad

Sorrel

LOW IN CALORIES AND FIBRE
EXCELLENT SOURCE OF VITAMINS A AND C
CONTAINS CALCIUM AND IRON

Origin

• Sorrel is a northern European salad green from the Polygonaceæ family that has long been popular from England to Russia. References to sorrel have been found in ancient Greek and Roman texts, and in medieval European manuscripts. During the 16th and 17th centuries in England, it was the main ingredient of *green sauce,* a widely used sweet-and-sour condiment that also contained marigold leaves and slices of citrus. By the 19th century, it was mainly sought after in France and middle European countries. Related to rhubarb, its name comes from the old French word for sour — "*surele*". Today, sorrel is cultivated in greenhouses in many parts of the world, including Canada and the U.S.

Varieties and description

• Its characteristic sharp, sour, lemony taste derives from its high content of oxalic acid. Its smooth, bright green leaves are shaped like large arrowheads, about 15 to 20 centimetres (6 to 8 inches) in length. Cultivated sorrel is milder tasting, sturdier, and larger than wild sorrel.

Availability

• Summer and fall are the main seasons for fresh sorrel, although it can be found year round when grown in greenhouses.

Selection

• Select bright green, firm leaves, although the sorrel will begin to wilt slightly just after picking.

Storage

• Use as soon as possible after purchasing. You can keep it a day or two in a plastic bag in the refrigerator if necessary.

Preparation

• Wash and dry carefully, then discard the stems by folding the leaf in half and holding in one hand, pulling the stem with the other hand. This strips away the stem all the way up the leaf. Use raw or cooked. Never cook sorrel in an aluminum or cast iron pot or it will discolour.

Suggestions

• Use a modest amount of raw sorrel in a salad, for an added piquancy.

• Sorrel sauce adds a great flavour to poultry dishes, egg dishes and many fish dishes.

• Sorrel soup is a well-known delicacy in some middle European countries as well as in Russia.

Nutritional Value

Sorrel
(per 100 g = 3½ oz)

			RDA**	%
calories*	(Kcal)	28		
protein*	(g)	2.1		
carbohydrates*	(g)	5.6		
fat*	(g)	0.3		
fibre	(g)	0.8	30	3
vitamin A	(UI)	12 900	3300	390
vitamin C	(mg)	119	60	198
thiamin	(mg)	0.09	0.8	11
riboflavin	(mg)	0.22	1.0	22
niacin	(mg)	0.5	14.4	4
sodium	(mg)	5	2500	0.2
potassium	(mg)	338	5000	6
calcium	(mg)	66	900	7
phosphorus	(mg)	41	900	5
iron	(mg)	1.6	14	11
water	(g)	90		

*RDA variable according to individual needs
**RDA : recommended dietary allowances (average daily amount for normal adult)

Sorrel salad

(serves 4)

Ingredients		
8	sorrel leaves	
3	endives	
125 g	mushrooms, sliced finely	1/4 lb
60-70 mL	toasted pine nuts	1/4 cup
	calendulas to garnish (optional)	

Dressing ingredients		
30 mL	walnut oil	2 TBSP
30 mL	balsamic vinegar	2 TBSP
2.5 mL	sugar	1/2 tsp
	salt and pepper to taste	

Instructions

- Finely chop four leaves of sorrel and leave the other four whole.
- Keep some endive leaves whole and chop the rest coarsely.
- Arrange sorrel, endives, mushrooms and pine nuts on a serving platter.
- Mix dressing and add just before serving.
- Garnish with flowers.

Purslane

EXCELLENT SOURCE OF VITAMIN A AND C
GOOD SOURCE OF IRON
SOURCE OF CALCIUM
VERY LOW IN FIBRE

Origin

• Originally found from eastern Europe to India, purslane, of the Portulaceæ family, is now beginning to regain its former popularity as a salad green. The plant spreads quickly, and can be found all over the world today. Thoreau extolled its virtues in his retreat at Walden Pond, writing: "I have made a satisfactory dinner... simply off a dish of purslane... which I gathered in my cornfield, boiled and salted."

Varieties and description

• Purslane is similar to watercress in appearance, although it has fewer leaves branching off each stem. Its tender, yellowish-green, delicate leaves are shaped like tear drops. The plant grows only about 5 or 6 centimetres (2 or 2 1/2 inches) in height, but spreads in large patches connected by common stems. The entire plant is edible, but usually only the leaves are sold. Purslane has a slightly tart taste, and is sometimes considered to be a herb.

Availability

• Purslane is generally available from June to September, and year round when grown in greenhouses.

Selection

• Choose purslane with firm stems, which indicate its freshness.

Storage

• Use as soon as possible.

Preparation

• Trim off the base of the stems, wash and dry. Chop into bite-sized pieces or leave whole.

Suggestions

• Use purslane in salads, especially green salads and tomato salads.

• Garnish a creamed soup with finely chopped purslane.

• Chop purslane and add to puréed vegetables or mashed potatoes just before serving for a zesty flavour.

• Purslane gives a good flavour to a stir-fry, but remember to add it at the end of cooking.

Nutritional Value

Purslane

(per 100 g = 3 1/2 oz)

			RDA**	%
calories*	(Kcal)	21		
protein*	(g)	1.7		
carbohydrates*	(g)	3.8		
fat*	(g)	0.4		
fibre	(g)	0.9	30	3
vitamin A	(UI)	2500	3300	76
vitamin C	(mg)	25	60	42
thiamin	(mg)	0.03	0.8	4
riboflavin	(mg)	0.1	1.0	10
niacin	(mg)	0.5	14.4	4
calcium	(mg)	103	900	11
phosphorus	(mg)	39	900	4
iron	(mg)	3.5	14	25
water	(g)	93		

*RDA variable according to individual needs
**RDA : recommended dietary allowances (average daily amount for normal adult)

Purslane and pear salad

(serves 2-4)

Ingredients		
70-125 g	purslane, left whole	1/8-1/4 lb
2	pears, peeled, cored and cut in quarters	
4	borage flowers for garnish	

Dressing ingredients		
45 mL	walnut oil	3 TBSP
30 mL	pear or other fruit vinegar	2 TBSP
	pinch of salt	

Instructions

- Mix all of the dressing ingredients together.
- Mix about 1/2 to 3/4 of the dressing with the pears to keep them from discolouring, and chill.
- When ready to serve, toss the remainder of the dressing with the purslane.
- Arrange the purslane and the pears decoratively on a serving platter and garnish with the borage flowers.

Chayote

(pronounced shy-o-tay)

SOURCE OF POTASSIUM, CALCIUM, PHOSPHORUS AND IRON
VERY LOW IN CALORIES, SODIUM AND FIBRE
CONTAINS VITAMINS A AND C

↑ Green chayote

Origin

• This versatile member of the Cucurbitaceæ (cucumber) family is also known as Christophene, Chu-Chu, Mirliton, Custard Marrow, Vegetable Pear, Chocho, Brionne and Pepinella. Chayote is native to Mexico and Central America, where the young squash leaves, flowers and the tuber-like roots are eaten in addition to the fruit. Closely related to squash and cucumber, it was once cultivated by the Aztecs, the Mayans and other native Americans. (Chayotl is the Nahuatl word for vegetable.) Today, one of the major exporter countries of the finest quality chayote is Costa Rica, but it is also grown all over the Caribbean and in many other parts of the world.

Varieties and description

• Shaped like a very large pear with a number of deep folds in its skin, the chayote's thin skin can be light green, white (blanco) or very dark green (negro), while inside its firm flesh is always white. It contains a soft flat seed that is edible after cooking. Chayotes have a crisp, fresh taste that is milder than most squash, and thus adapts well to many recipes.

Availability

• Available all year round, although more plentiful in the winter.

Selection

• Look for a chayote that is very firm, free of blemishes, and shows no signs of aging.

Storage

• Wrapped in plastic, it can be refrigerated for several weeks.

Preparation

• Although the peel of chayotes is edible, they sometimes have a few prickly spines on the surface and are better peeled. When peeling chayotes, your hands may become coated by their juices; so use gloves or peel under running water. Chayote has a wonderful crispy texture, and a very mild flavour that blends well with almost every herb and spice. It can be prepared in a great variety of ways.

Suggestions

• Slice a chayote and sprinkle with lemon juice and salt.

• Add it to any salad for a crispy texture, such as tuna, chicken or vegetable salad.

• Substitute it for other types of summer squash in your favourite recipes.

• Cut a chayote in chunks, then boil or steam until tender (about 10 to 15 minutes) and serve with butter, salt and pepper. You can also add grated cheese.

• Try it in soups, puréed or diced.

• Stuff chayotes with rice and vegetables, savory meat or a seafood mixture and bake au gratin.

• Chayotes are an excellent addition to chutney.

Chayote and red kidney bean salad

(serves 4)

Ingredients		
2	chayotes	
2	scallions (green onions), minced finely	
250 g	red kidney beans, cooked	1/2 lb
8	cherry tomatoes	
	Boston lettuce	

Dressing ingredients		
45 mL	salad oil	3 TBSP
	juice from 1 lime	
5 mL	fresh thyme, minced finely	1 tsp
	salt and freshly ground black pepper to taste	

Instructions

- Peel the chayote, cut in half and slice finely.
- Cook chayote slices in boiling water from 3 to 5 minutes, until just tender. Pour cold water over them and drain. (Leave raw if preferred.)
- Arrange chayote mixed with scallions and the beans on a serving platter, or gently mix it all together in a salad bowl.
- Mix all of the dressing ingredients together.
- Pour dressing over ingredients, and let marinate at room temperature for 30 minutes.
- Serve on a bed of lettuce and decorate with cherry tomatoes.

Negro ↑

↓ Blanco

L'Exotic's chayote and ginger salad

(serves 4)

from the Montreal restaurant L'Exotic

Ingredients		
2	chayotes, peeled and sliced thinly	
30 mL	fresh parsley, finely chopped	2 TBSP
30 mL	scallions (green onions), finely chopped	2 TBSP
30 mL	light salad oil (such as sunflower seed oil)	2 TBSP
	juice of half a lemon	
5 mL	fresh ginger, finely grated	1 tsp
	salt and pepper to taste	

Instructions

- Toss all the ingredients together.
- Chill.
- Covered, it will stay fresh and delicious tasting for several days.

Nutritional Value

Chayote

(1 med. half chayote = 100 g = $3^{1}/_{2}$ oz)

			RDA**	%
calories*	(Kcal)	28		
protein*	(g)	0.6		
carbohydrates*	(g)	7.1		
fat*	(g)	0.1		
fibre	(g)	0.7	30	2.3
vitamin A	(UI)	20	3300	0.6
vitamin C	(mg)	19	60	32
thiamin	(mg)	0.03	0.8	3.8
riboflavin	(mg)	0.03	1.0	3
niacin	(mg)	0.4	14.4	3
sodium	(mg)	5	2500	0.2
potassium	(mg)	102	5000	2
calcium	(mg)	13	900	2
phosphorus	(mg)	26	900	3
iron	(mg)	0.5	14	4
water	(g)	90		

*RDA variable according to individual needs
**RDA : recommended dietary allowances (average daily amount for normal adult)

Fresh herbs

PARSLEY IS A GOOD SOURCE OF
VITAMINS A AND C
RICH IN IRON
LOW IN CALORIES

Origin

• Herbs have been around since cooking was invented. Ancient manuscripts tell us that the Egyptians, Romans, Greeks, Persians and Chinese, among others, both cultivated and picked wild herbs extensively for medicinal purposes, religious rituals, to preserve foods (by pickling, marination, etc.) long before refrigeration was invented, to flavour foods and even to use as aphrodisiacs. It is the flavour given by indigenous herbs and spices that make regional cuisines so special. Now that fresh herbs are available on the market (many of them grown hydroponically), you can recreate the aromatic flavours of delicious dishes from anywhere in the world.

Varieties

• **Sweet basil, lemon basil** and **red basil** — Among the most perishable and delicious herbs, basil is excellent in salads, pasta, soups, egg dishes, poultry dishes and gravies. Fresh basil is fabulous with tomatoes, garlic and olive oil, in potato salad, or in bean salads.

• **Chervil** — One of the celebrated French "fines herbes", chervil looks similar to parsley, but has an elusive hint of licorice. Use chervil as you would parsley, especially in egg dishes, for an extra gourmet taste.

• **Tarragon** — One of the strongest herbs, tarragon should be used in moderation as it will overpower other flavours. Best used in conjunction with other strong herbs such as chives, and with lemon, garlic and parsley, tarragon is often used in poultry or fish dishes and in sauces.

• **Bayleaf** — Essential to Spanish, Italian and other Mediterranean cuisines, bayleaves are excellent in tomato and meat sauces and are frequently used to flavour soups, stews, marinades and pickling brine.

• **Marjoram** — Like oregano though slightly milder, marjoram compliments most foods, particularly omelets and other egg dishes, lamb, chicken and any dish with tomatoes.

• **Oregano** — It is impossible to think of Italian food without oregano. It compliments all Mediterranean vegetable, fish, chicken and meat dishes, as well as dishes with cheese, shellfish or pasta.

• **Parsley** (curly or Italian) — The most versatile and most easily obtainable herb, parsley blends well with other herbs and goes well with any dish. Try a simple pasta topping made with lots of chopped parsley, minced and lightly browned garlic, olive oil, Parmesan cheese, salt and pepper.

• **Sage** — There are many varieties of sage and they are all excellent in stuffings (especially for poultry and lamb), sausages and poultry dishes. Sage is very good combined with celery and onions.

• **Savory** — This herb makes an attractive garnish. Its piquant flavour is excellent for herb butter and in most soups and bean or pea dishes. Try it in sauces and salad dressings.

• **Thyme** — More than fifty varieties of thyme are grown, though few are well known. It's excellent in all Mediterranean dishes, in salad dressings, soups, on roast meats, poultry and grilled fish, in casseroles and with many vegetables. Thyme mixes particularly well with rosemary, oregano and sage. Use in any dish cooked with red wine. Lemon thyme is excellent in stuffings.

• **Rosemary** — One of the more familiar herbs, rosemary is a perfect flavouring for roasted dishes, herb breads, soups, sauces, dressings and all Mediterranean dishes.

• **Mint** — One of the few herbs that compliments both fruit and vegetables and both savory and sweet dishes, mint is delicious with new potatoes, peas and beans, in sauces and salad dressings (especially with garlic and yogurt), with lamb and veal, and in Arabic and Indian cuisine. Mint is also good in desserts with chocolate or fresh fruit such as melon or oranges. Also try spearmint, pineapple mint, orange mint and peppermint. Mint is a terrific garnish for strawberries.

• **Dill** — Popular in Scandinavian, Russian and Middle European cuisines, dill can add a special taste to your cooking. It is wonderful in soups, salads, fish dishes, poultry dishes and many others. It is delicious in cucumber salads and sprinkled on sour cream served with baked potatoes. Try a dill dressing with avocado, tomato and onion salad.

• **Coriander** — Also called **cilantro** or **Chinese parsley,** coriander is an essential herb for Latin American, Chinese, Indian and Southeast Asian cooking. Use it in salsa, guacamole, soups, sauces, oriental stir-fries and curries.

• **Chives** — One of the mildest members of the onion family, chives give a subtle oniony taste to a variety of foods, such as sour cream and cream cheese (for dips), eggs, potato dishes, poultry, fish, salad dressing, sauces, soups and many others. It makes an attractive garnish.

• **Anise** — Like mint, anise can be used in either sweet or savory dishes. It is tasty in salads and soups, and adds an interesting flavour to many cooked vegetables. It's delicious in fish and poultry dishes. It is often used in Arabic and Indian cooking.

Availability

• Fresh herbs are available year round.

Selection

• Generally, herbs that look fresh are fresh, especially such fine-leafed herbs as dill and parsley. Avoid dark spots on such larger leafed herbs as basil and mint.

Storage

• Herbs are fresh green plants and should be refrigerated and used within a few days of purchase. Wrapped in a slightly damp paper towel, the hardier herbs will last much longer than the fine-leafed ones.

• Many of the fine-leafed types, such as dill, chervil, chives, basil, parsley and cilantro, can be chopped very finely and frozen. Many others, such as rosemary, oregano and thyme, can be dried for later use: Tie a thread around the base and hang them upside down. Herbs lose the intensity of their flavour when they are dried and are always better-tasting when used fresh.

Preparation

• As a rule, use double or triple the amount of fresh herbs to substitute for dry herbs in your favourite recipes. Herbs can be prepared several ways, or left whole for garnishing or for flavouring dishes if you want to remove them easily after cooking (eg. bay leaves, mint fresh bouquet garni). Some, such as chives and dill, are easily snipped with a scissors. Most can be minced with a knife. Some, such as basil or sage, can be torn into bite-sized pieces by hand. Discard any woody or fibrous stems before preparation (eg. rosemary or oregano stems). Fresh herbs that are grown hydroponically generally don't need washing, while those sold loose or in small tied bunches (usually parsley, cilantro, basil, etc.) need to be washed and dried.

Suggestions

• Herbs make an excellent, flavourful substitute for salt.

• For an extra good salad dressing, add about 45 to 60 mL (3-4 TBSP) of your favourite fresh herbs to a litre of wine vinegar or cider vinegar and let it steep for a week or so. Discard the herbs and store the flavoured vinegar in a cool, dark place.

• Herb-flavoured butter is delicious served on steamed or boiled vegetables, grilled fish, steak or with pasta. Simply blend the minced herbs into softened butter and refrigerate.

• Even a simple snack like cheese melted on toast can be improved enormously by sprinkling on fresh herbs (such as rosemary, oregano or thyme) before grilling.

• Make your own bouquet garni with fresh parsley, thyme and a bay leaf to add to soups, casseroles and stews. Tie the herbs with a thread or in a bit of cheesecloth, then discard before serving. Celery leaves, whole garlic and orange peel can add extra flavour.

Nutritional Value

Parsley

(per 10 g = 1/3 oz)

			RDA**	%
calories*	(kcal)	4		
protein*	(g)	0.5		
carbohydrates*	(g)	0.8		
fat*	(g)			
fibre	(g)		30	
vitamin A	(IU)	850	3300	25.8
vitamin C	(mg)	17	60	28.3
thiamin	(mg)	0.01	0.8	1.3
riboflavin	(mg)	0.02	1.0	2
niacin	(mg)	0.1	14.4	0.7
sodium	(mg)	4	2500	0.2
potassium	(mg)	73	5000	1.5
calcium	(mg)	20	900	2.2
phosphorus	(mg)	6	900	0.7
magnesium	(mg)	4	250	1.6
iron	(mg)	0.6	14	4.3
water	(g)	85		

*RDA variable according to individual needs
**RDA recommended dietary allowance (average daily amount for a normal adult)

1. Marjoram 2. Dill 3. Bay leaf 4. Rosemary 5. Sweet basil 6. Chervil 7. Oregano
8. Chives 9. Tarragon 10. Coriander 11. Thyme 12. Sage 13. Mint

Tabouli

(serves 4 - 6)

Ingredients

250 mL	uncooked bulgar*	1 cup
250 mL	vegetable or chicken stock, hot	1 cup
60 mL	finely minced parsley	1/4 cup
15-30 mL	finely minced coriander	1-2 TBSP
60 mL	finely minced mint	1/4 cup
125 mL	finely minced scallions	1/2 cup

2	medium tomatoes, peeled and finely chopped	
1	medium cucumber, peeled, seeded and finely chopped	
60 mL	olive oil	4 TBSP
75 mL	lemon juice	1/3 cup
	salt and pepper to taste	

Instructions

- In a bowl, add the bulgar* to the hot stock (or hot water). Cover and let sit for half an hour, or until liquid is absorbed.
- Add the rest of the ingredients, except the salt and pepper, and mix well. Cover the bowl and leave for an hour or two for flavours to blend.
- Season to taste and serve.

** Also known as cracked wheat*

Cantaloupe

EXCELLENT SOURCE OF VITAMINS A AND C
CONTAINS POTASSIUM AND MAGNESIUM
SOURCE OF FIBRE
VERY LOW IN CALORIES

Origin

• A true cantaloupe — a coarse, scaly, green-rinded melon — is a European melon that is never seen in North America. Cantaloupes were named after Cantalupo, a summer residence of the popes in the Middle Ages where Armenian melons were grown. Some sources say that Pope Paul II was so fond of melons that he died from overeating them.

• The melon that we call cantaloupe is a kind of musk melon, or aromatic melon, that originated in the vast area from eastern Turkey to Persia through to Afghanistan and northwestern India, as well as the Russian territories to the north that border this entire region. The earliest records show that man began melon cultivation in India and Africa. Melons were introduced to the Roman Empire about 2,000 years ago, and to China about 800 years later. Columbus and other explorers brought seeds to North America, and the Indians and early settlers spread melon cultivation throughout the continent. Jacques Cartier noted in 1535 that Indians in the Montreal area were growing melons.

• Today, the major producer countries of cantaloupes for our market are the United States, Mexico, Guatemala, Honduras and the Dominican Republic. Canada produces some cantaloupes for local consumption.

Varieties and description

• Usually oval in shape, although some varieties are elongated, the cantaloupe has a pronounced tan coloured, fibrous, cork-like netting raised over a golden, yellowish-green background. Some varieties have visible grooves that look like segment markers. The melon rind is normally hard, and often shows tiny cracks near the stem end, which is slightly depressed. They are about 15 centimetres (6 inches) or larger in diametre, although some varieties may be smaller. Most cantaloupes have a lush orange flesh containing a small, moist clump of inedible seeds in the centre, though a few varieties of green-fleshed cantaloupes have been developed. Cantaloupes are sweet, refreshingly juicy and wonderfully fragrant when ripe.

Availability

• Because so many countries grow cantaloupes, they are available year round. June, July and August are the peak months.

Honeydew

VERY LOW IN CALORIES
EXCELLENT SOURCE OF VITAMIN C
TRACES OF CALCIUM AND IRON
LOW IN FIBRE

Origin

- All melons are members of the Cucurbitaceæ (cucumber) family and have the same initial origins. Melons cross breed easily and new varieties are always coming on the market. Green honeydew is grown extensively in the U.S., Mexico, throughout Central America and in the Dominican Republic. Orange honeydew is grown mainly in California and the Dominican Republic.

Varieties and description

- Green honeydew, which usually weighs between 1.75 and 3.5 kilos (3 1/2 and 7 pounds), has a smooth, creamy white or pale yellow rind slightly tinted with pale green, which turns an even creamier colour as it ripens. The melon shape is slightly elongated, and its surface seems to become dull and waxy as it ripens. Its attractive green flesh is very juicy and as sweet as honey. Orange honeydew is much smaller, and is more like a cantaloupe in size, taste, texture and smell, as well as colour. Its smooth white rind turns a very pale coral pink when fully ripe. All honeydews have very hard rinds that give a little when ripe.

Availability

- Green honeydew is available all year, though it is most plentiful from June to October. Orange honeydew from California is in season from July to October, while those from the Dominican Republic can be found from December to June.

Selection

- Avoid very green melons, as they will never ripen, and melons with soft spots.

Storage

- Unless your melon is ripe and ready to eat, keep it at room temperature until a fruity aroma develops, then in the refrigerator for up to a few days. It tastes best when eaten as soon as it's ripe, served at room temperature or only lightly chilled. All melons should be wrapped in plastic in the refrigerator.

Preparation

- Simply cut in wedges or in half, discard seeds from the portion you are using, and serve.

Suggestions

- Try creating a fruit basket from your melon, cutting a wedge from each side of the top that will leave a basket "handle" in the middle, or cut in half and flute the edges. Scoop out the melon flesh and fill the basket with strawberries, pieces of melon and grapes.
- A dash of ginger marinated in some lemon juice adds zest to a very sweet honeydew.
- A little fresh mint compliments a melon-based fruit salad.

Nutritional Value
Honeydew
(per 100 g = 3 1/2 oz)

			RDA**	%
calories*	(kcal)	33		
protein*	(g)	0.8		
carbohydrates*	(g)	7.7		
fat*	(g)	0.3		
fibre	(g)	0.6	30	2
vitamin A	(IU)	40	3300	1.2
vitamin C	(mg)	23	60	38.3
thiamin	(mg)	0.04	0.8	5
riboflavin	(mg)	0.03	1.0	3
niacin	(mg)	0.6	14.4	4.2
sodium	(mg)	12	2500	0.5
potassium	(mg)	251	5000	5
calcium	(mg)	14	900	1.6
phosphorus	(mg)	16	900	1.8
magnesium	(mg)		250	
iron	(mg)	0.4	14	2.9
water	(g)	89		

* RDA variable according to individual needs
** RDA: recommended dietary allowance (average daily amount for a normal adult)

Melon nectar

(serves 2)

Ingredients

1	honeydew or other melon
2	cinnamon sticks
1	dozen balls of another kind of melon

Instructions

- Cut the melon in half, discard seeds, scoop out the flesh and liquefy it in a blender.
- Pour the liquid back into each melon half, place a cinnamon stick in each one, and refrigerate for at least 20 minutes.
- Provide a drinking straw with each melon half.

Japanese melons

(Tokyo King / Emerald Jewel / Sapphire / Amur)

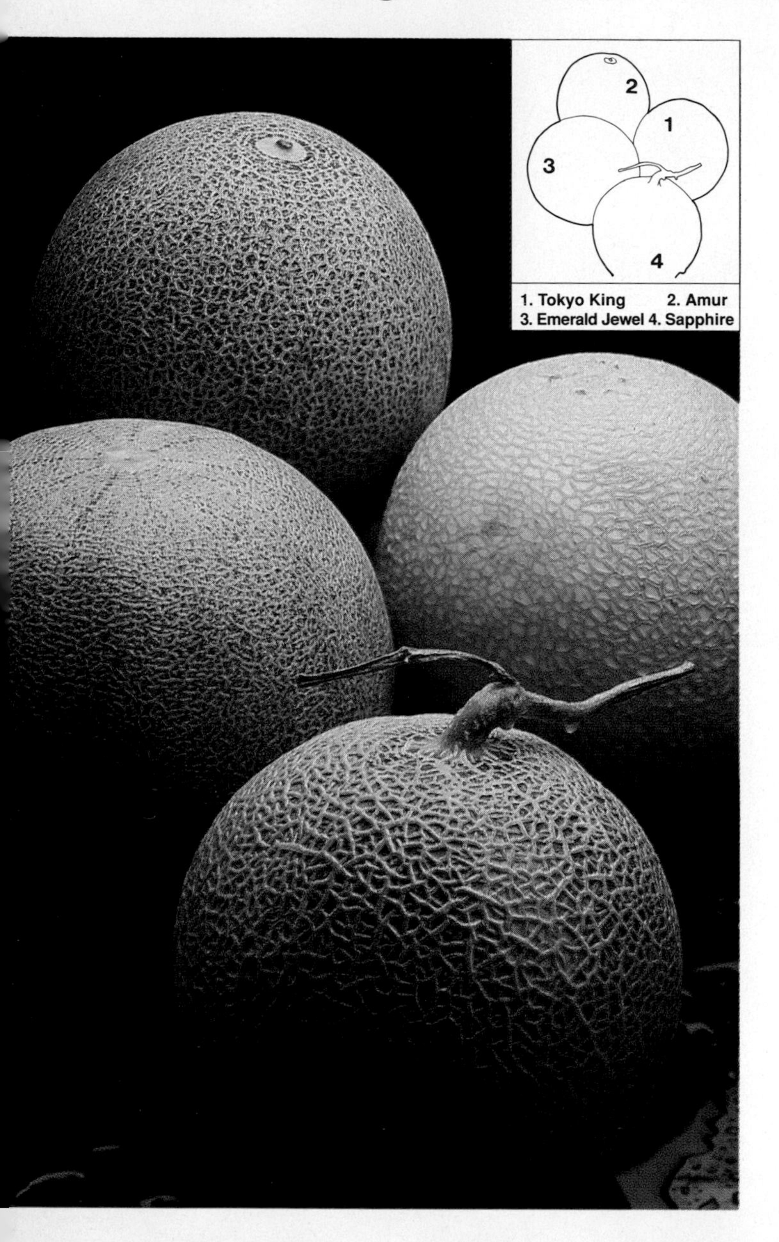

Origin

• These melons are hybrids perfected in Japan since the Second World War. A number of growers in Oregon cultivate these delectable melons for the North American market. The seed is still imported each year from Japan and is sometimes in short supply. These melons are usually presented as special gifts in Japan, and in Tokyo's most exclusive fruit store can cost up to $150 U.S.!

Varieties and description

• Tokyo King is the older, more proven variety, with a beautifully netted, pale cream skin. Its flesh is also cream-coloured and very sweet. It usually weighs between 1.5 and 2 kilos (3 and 4 pounds).

• Emerald Jewel is round to oval in shape, with a well-netted dark green skin and light green flesh. It, too, is sweet and it is about the same size as the Tokyo King. The seed cavity is very small, so the flesh is quite thick.

• Sapphire melons are also roughly the same size as Tokyo Kings, but they have a larger netting pattern.

• An Amur melon is similar to an Emerald Jewel in size and appearance, but its netting is much finer. It also has thick, light green flesh that is slightly less sweet than the other Japanese dessert melons.

Availability

• These exquisite melons have a short season that runs from mid-July to mid-August.

Selection

• Shipped ripe and ready to eat, with foam netting to protect them from bruising, these melons should be in excellent condition.

Storage

• Although you can keep them refrigerated for a few days, these special melons taste best eaten as soon as possible.

Preparation

• As with other melons, simply cut open, remove seeds, and serve.

Suggestions

• Any one of these melons makes a simple and delicious dessert.

• Offer them as gifts.

Specialty melons

(Crenshaw/ Persian / Casaba / Spanish)

Varieties and description

• A good crenshaw or cranshaw melon is probably the most delicious melon in the world. Its smooth dark green rind should turn golden yellow when ripe, and its flesh is a pale salmon pink. It is round in shape with a pointed stem end, weighs between 2 and 3.5 kilos (4 and 7 pounds), and exudes a wonderful smell when ripe. Crenshaw is a very fragile melon and should be handled with care.

• Persian melons are similar to very large cantaloupes in appearance except that they have no grooves or segment markers. Usually from 3 to 3.5 kilos (6 to 7 pounds) in weight, they are difficult melons to grow, and very fragile when ripe. It isn't hard to tell when a Persian melon is ripe, as its skin turns more golden, its netting lightens in colour, its perfume becomes very pronounced and its thin skin softens a little. The flesh is a luscious deep orange.

• A casaba melon has a yellowish furrowed or wrinkled skin that tends toward green near the stem end. It weighs about 2 to 3.5 kilos (4 to 7 pounds), its shape is round and one end is pointed. Its flesh is a creamy white colour tinged with pale yellow. Casaba melons are not very aromatic unless ripened on the vine. The only sure indications of ripeness are a slight stickiness and a slight softening.

• A Spanish melon has a dark green, hard, grooved skin that does not soften or change colour with ripening. The aroma is not pronounced, making it difficult to determine when to eat it. It tends to get slightly sticky when ripe, which will be your best clue. It is similar in shape and size to a casaba, but its juicy flesh is a pinkish yellow. One variety of Spanish melon is the tendral melon, which also has dark green, grooved skin. Its flesh is greenish yellow.

Availability

• Spanish and tendral melons are brought in from Spain in November and December. The crenshaw, Persian and casaba melons grown for our market arrive mostly from California between July and October.

Selection

• Look for a well shaped, unbruised melon. Don't depend on squeezing the stem end, or you may be feeling the softness resulting from other customers' squeezes! There is no infallible method for choosing a great melon, as it always depends on the melon being harvested at the precise point when it has all its sugars, yet is not too ripe or it would be damaged by transportation from field to market to store.

Storage

• Keep melons at room temperature until ripe, then use, or refrigerate for a day or two. For the best flavour, use as soon as the melon ripens. Wrap melons in the refrigerator to prevent their ethylene gas from affecting other foods.

Preparation

• Cut in half or cut wedges from the melon, removing the seeds only from the portion you are going to eat. Melons are best when eaten at room temperature or only lightly chilled.

Suggestions

• Crenshaw melon is so delicious that it deserves to be eaten in its natural state.

• Persian melons are good in fruit salads, grilled in kebabs, served with ice cream or eaten as is.

• Casaba melons are very sweet, but don't have as strong a flavour as other melons. They are good with a little lemon or lime juice sprinkled over them, or combined with other fruits in a fruit salad.

• Spanish and tendral melons are often served with dried meats like prosciutto or smoked fish. Cut in balls or cubes and sprinkle with port or sherry to use as an hors d'oeuvre, or sprinkle with midori (a sweet liqueur with a strong honeydew taste) and use as a quick, refreshing dessert.

• After a heavy or spicy meal, for a great dessert, mix some finely chopped fresh mint into plain yogurt, sweeten a little if necessary, and serve as a dip for pieces of melon.

1
1
2
2
3
3
4
5
5
1. Crenshaw
2. Santa Claus
3. Casaba
4. Juan Canary
5. Sharlyn

Melon with cinnamon syrup

(serves 4)

Ingredients		
500 mL	fine sugar	2 cups
250 mL	orange juice	1 cup
250 mL	pineapple juice	1 cup
250 mL	grapefruit juice	1 cup
250 mL	apple juice	1 cup
60 mL	kirsh	4 TBSP
60 mL	melon liqueur	4 TBSP
1	cinnamon stick	
1	melon, Casaba or Spanish	
4-6	mandarins	

Instructions

- Combine first 7 ingredients in a saucepan and bring to a boil.
- Let simmer 20 minutes, stirring occasionally.
- Remove from heat, add cinnamon stick, cool, and refrigerate for several hours.
- Cut the melon into ball shapes, and the mandarins into segments, removing any pips.
- Pour the syrup over the fruit and serve. Keep any leftover syrup for your next fruit salad.

Nutritional Value
Melons

Most orange melons have nutritional values comparable to cantaloupe.

Most green and very pale melons have nutritional values comparable to honeydew.

Other specialty melons

(Juan Canary / Santa Claus / Sharlyn / Ogen / Galia)

Varieties and description

• Juan Canary melon has a smooth, beautiful canary yellow rind and sweet white flesh that is slightly pinkish around the seed cavity. These melons are elongated and are usually very aromatic when ripe.

• A Santa Claus, or Christmas melon, is similar in appearance to a small, elongated watermelon. The rind is a mottled golden yellow, green and black, and its flesh, reminiscent of honeydew, is a pale green.

• Sharlyn melons are elongated, and have a slight beige netting over a rind that turns from green to orange. Its yellowish-orange flesh is extremely sweet and aromatic.

↓ Galia and Juan Canary melons

• Ogen are named after the kibbutz in Israel that first developed them about 30 years ago. These are small melons, with a smooth, segmented, greenish-yellow rind, and very juicy, pale green flesh. In addition to Israel, Spain and France now cultivate ogen melons.

• Galia is another Israeli hybrid melon. Named after the daughter of its developer, Galia melons are netted, segmented, yellowish-green on the outside, and pale green inside. Some smaller Galias have bright yellow rinds that show no segmenting. This aromatic, delicious melon has become very popular, and is now cultivated in other countries, including Spain and Italy.

Availability

• Juan Canary melons come from Brazil between December and March, and from the U.S. between July and October. The season for American Santa Claus melons is July to October. Sharlyns come from the United States during the summer. Ogen and Galia melons are available from April to December.

Selection

• It isn't always easy to choose a good melon. When in doubt, ask the advice of a store's produce manager. If a melon is picked after its sugar content has developed to the full extent, and if it has arrived at the store in good condition, it should be great.

Storage

• Do not refrigerate your melon until it has fully ripened, then use it quickly or refrigerate it for a few days. Remember that during the summer, melons often arrive in a ripened state, and will not benefit from sitting on your counter.

Nutritional Value
Melons

Most orange melons have nutritional values comparable to cantaloupe.

Most green and very pale melons have nutritional values comparable to honeydew.

1
1
2
2
3
4
4
3
1. Honeydew 2. Persian melon
3. Orange fleshed Honeydew
4. Cantaloupe

↓ **Prince melon**

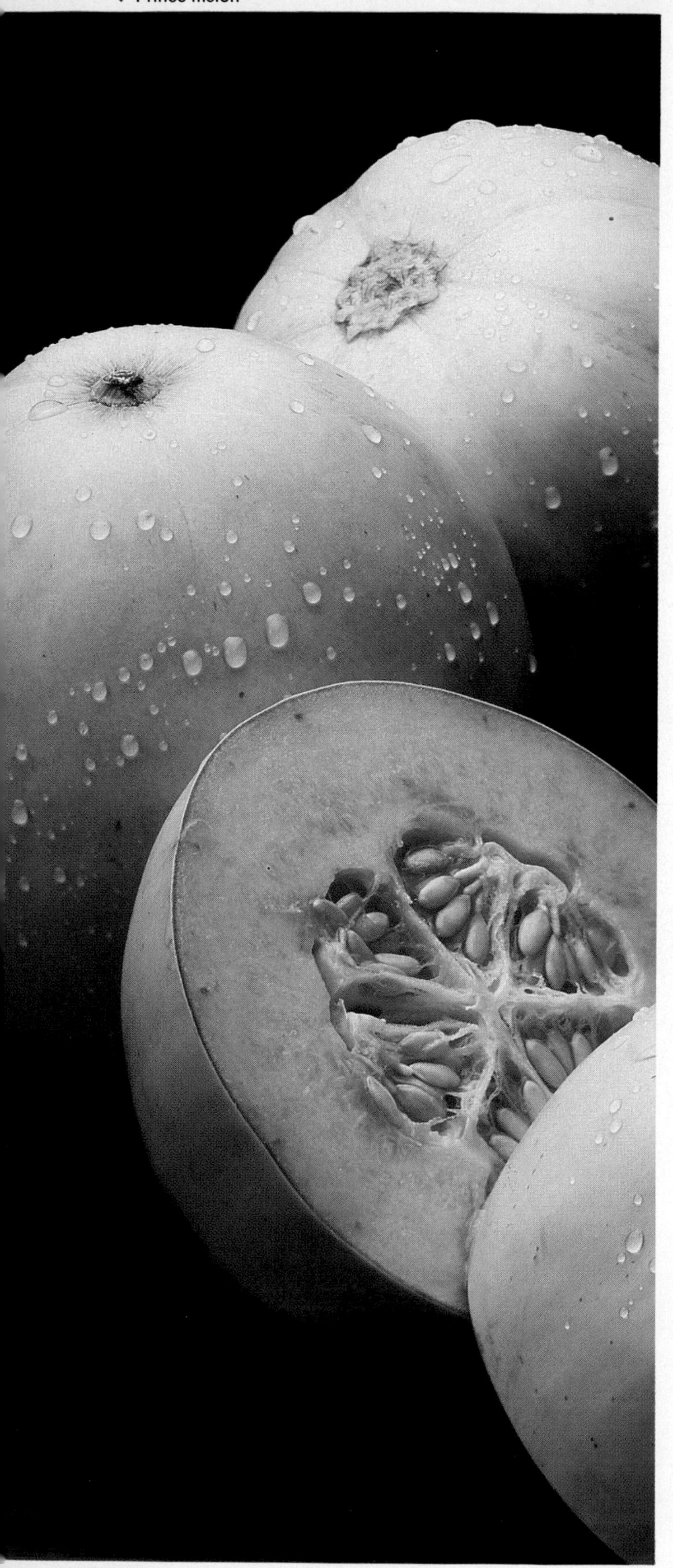

Preparation

- Cut in half, or in wedges, remove seeds, and serve.

Suggestions

- Hollow out half a large watermelon and fill it with melon balls of different colours for a party centrepiece. Sprinkle with blueberries for a beautiful contrast of colour and flavour.
- Scoop the flesh out of a melon in ball shapes. Turn the halves upside down on a serving platter, and attach the melon balls to the rind with toothpicks. You can add strawberries, grapes, etc., attached the same way, to create a festive dessert.

↓ **Mayan melon**

• For a luscious treat, make fruit kebabs, alternating chunks of melon, pineapple, banana, grapes, and other fruits on kebab sticks, and drizzle a little lemon juice mixed with honey over them.

• Use melons to make a delicious chutney to serve with a wide variety of meals. Cube 2 or 3 melons (of different varieties, if desired), and place in large saucepan with the following ingredients: 2-3 diced onions, 2-3 crushed garlic cloves, 500 mL (2 cups) each brown sugar and white wine vinegar, 30 mL (2 TBSP) fresh grated ginger, 5 mL (1 tsp) each ground cumin seeds, cloves, cinnamon and salt, and a good dash of nutmeg and cayenne. Add a hot pepper, if desired. Boil, then cook on medium low for 30 minutes. Add 125 mL (1/2 cup) raisins, then continue cooking for 30 minutes or more, stirring more frequently, until the liquid thickens. (Be careful not to crush the pieces of melon when stirring.) Adjust seasoning, cool, put in jars and refrigerate.

Melon Sorbet

(serves 4)

Ingredients		
190 mL	sugar	3/4 cup
250 mL	water	1 cup
560 mL	puréed melon	2 1/4 cups
30 mL	lemon juice	2 TBSP
15 mL	grated cucumber	1 TBSP

Instructions

• Bring the sugar and water to a boil and let simmer about 3 minutes.

• Remove from the heat, add the rest of the ingredients.

• Let cool at room temperature, then freeze, stirring from time to time to ensure a uniform texture.

• For a beautiful presentation, make 2 half recipes using two different coloured melons, and serve everyone a scoop of each.

Pepino

Origin

• Also called melon pear, treemelon, mellow fruit, or sweet cucumber, this unusual melon of the cucumber family, originated in Peru. It is now grown in several South American countries, including Chile, as well as in California, but most of the pepino destined for our market is imported from New Zealand.

Varieties and description

• Averaging 10 to 12 centimetres (4 to 5 inches) in length, this unusual oval fruit looks more like an exotic eggplant than a melon. It has a very thin, glossy skin, which ripens from a pale green to pale, creamy yellow with purple stripes. The ripe fruit gives off a delicious aroma. Its firm, slightly sweet, melon textured flesh is yellow. Pepinos contain a small clump of soft edible seeds at its core. Milder tasting than most melons, it is an ideal size for one person.

Availability

• New Zealand pepinos are available from January to October, with peak supplies from March through June. California pepinos, which are not often seen in Canada, are in season from August to December.

Selection

• Pepinos should feel very firm, and have a smooth, unblemished skin. They are wrapped individually for protection.

Storage

• Keep your pepino at room temperature until the skin changes colour to yellow with vivid purple streaks. When ripe, they are best eaten right away, though you can keep them refrigerated for a day or two wrapped in a plastic bag.

Preparation

• Just peel and slice it, or cut in half and scoop out the flesh with a spoon. The seed section is sweet and edible.

Suggestions

• Eat a pepino by itself, with a squeeze of lemon or lime to bring out its flavour, for breakfast, a snack or for dessert.

• Add it to a fresh fruit salad.

• Try cutting a pepino in cubes, then sprinkle on a spoon of sugar, a squeeze of lemon, finely minced crystallized ginger, and a dash of Cointreau or Grand Marnier. Chill and serve.

• Halve it, scoop out the centre and dice it, then mix it with chicken or turkey salad, shrimp or seafood salad, or fruit salad. Fill the pepino shells and garnish with a twist of lemon.

• Left whole, pepinos are very attractive in a centre piece or fruit bowl.

Pepino and pear cup

(serves 4)

Ingredients		
2	pepinos	
2	pears	
15 mL	fresh mint, finely chopped	1 TBSP
15-30 mL	lemon juice	1-2 TBSP
30 mL	brown sugar	2 TBSP

Instructions

- Peel the pepinos, cut in half, discard seeds, and cut into bite-sized pieces.
- Use Asian apple-pears if possible. Core and slice.
- Combine all the ingredients, mix thoroughly, and refrigerate for an hour or two.
- Serve in glass goblets with vanilla or chocolate ice cream or tofutti.

Watermelons

(Mini /seedless / yellow)

LOW IN CALORIES
HIGH IN VITAMIN C
SOURCE OF VITAMIN A AND THIAMIN

Origin

• Watermelon probably originated in Africa, where it was found growing in the wild by European explorers in semi-tropical areas, as well as in the Kalahari desert. Native peoples used watermelon as an important source of water for themselves and for their animals. According to the first French explorers of the Mississippi Valley region, Indians used to cultivate them, leading some botanists to believe that they might also be native to North America. Today, watermelons are a popular treat throughout the world, and growers are beginning to concentrate more on specialty watermelons to please the North American public. To many consumers' delight, the Japanese have developed a seedless watermelon, which is now being cultivated in California, Florida and Georgia. Yellow watermelons are imported from Taiwan, Mexico and Florida. The best mini watermelons are grown in Guatemala and Costa Rica.

Varieties and description

• The Central American mini watermelons are round, red-fleshed and sweet. They usually weigh about 4 to 5 kilos (8 to 10 pounds). Seedless watermelons are usually round in shape and weigh about 4 to 5 kilos (8 to 10 pounds). The striped rind is rather thin, but strong, and the crisp, sweet, juicy flesh is an attractive deep red colour. Seedless watermelons may contain a few small, white, edible seeds. They do not, of course, produce seeds for the next crop and the process involved in creating seeds is very complex and costly. Yellow watermelons from Taiwan, which are the best tasting and have the prettiest colour, are called Jade watermelon. The taste is not very different from that of the red varieties.

Availability

• The season for the usual types of watermelon runs from March to October, with the peak months from June to August. Seedless watermelons can be found on the market from May to October. The Central American mini watermelons have two seasons, from March to June and from September to November. The yellow watermelons from Taiwan are usually available from October to December, while those from Mexico come in from February to April. The Florida yellow watermelons are more of a summer variety.

Selection

• Most experts agree that the only foolproof way to ascertain the ripeness and taste of a watermelon is to cut it open. A keen ear may be able to hear how ripe a watermelon is, when tapped; a heavy thud indicates ripeness, while a hollow sound means that the fruit is not yet ready.

Storage

• Watermelons are best stored in the refrigerator, wrapped in plastic. The seedless variety has a longer shelf life than other kinds, because of the lack of seeds. The seeds are the producers of the natural ethylene gas that hastens the ripening process of melons.

Preparation

• Mark Twain once said of watermelon: "When one has tasted it, he knows what angels eat". Cut in rounds, then halve or quarter for easy eating. Serve cold.

Suggestions

• From picnics to parties, watermelon is the perfect summer treat.

- Watermelon combines well with other fruit for fruit salad.
- Try pickling the rind of watermelon.
- For a summer party, hollow out a watermelon half and serve a fruit punch in it.

Striped melon sorbet

(serves 6-8)

Ingredients		
1	packet unflavoured gelatin	
375 mL	water	1 1/2 cups
500 mL	watermelon, cubed	2 cups
500 mL	honeydew, cubed	2 cups
10 mL	fructose	2 tsp
190 mL	sugar	3/4 cup
60 mL	fresh lemon juice	1/4 cup
	fresh mint for garnish	

Instructions

- Dilute the gelatin in 125 mL (1/2 cup) cold water, heat until the gelatin is dissolved, separate mixture in 2 equal quantities, and set aside until it becomes lukewarm.
- In a blender, combine the watermelon (after removing seeds), 125 mL (1/2 cup) water, and half of the fructose, the sugar, and the lemon juice. Mix in half of the gelatin. Purée.
- Repeat the last instruction, substituting honeydew for watermelon, and using the rest of the ingredients.
- Put the two mixtures into 2 different bowls, and place them in your freezer.
- After a few hours, before the liquids freeze solid, reprocess them in the blender, then refreeze.
- Serve in parfait glasses, alternating the two different coloured sorbets to create a striped effect. Decorate with a sprig of fresh mint, and serve with fresh fruit.

Nutritional Value

Watermelon

(per 160 g = 5 1/2 oz)

			RDA**	%
calories*	(kcal)	50		
protein*	(g)	1.0		
carbohydrates*	(g)	11.5		
fat*	(g)	0.7		
fibre	(g)	0.5	30	1.7
vitamin A	(IU)	585	3300	17.7
vitamin C	(mg)	15	60	25
thiamin	(mg)	0.13	0.8	16.3
riboflavin	(mg)	0.03	1.0	3
niacin	(mg)	0.3	14.4	2
sodium	(mg)	3	2500	0.1
potassium	(mg)	186	5000	3.7
calcium	(mg)	13	900	1.4
phosphorus	(mg)	14	900	1.5
magnesium	(mg)	17	250	6.8
iron	(mg)	0.8	14	5.7
water	(g)	146.4		

*RDA variable according to individual needs
**RDA: recommended dietary allowance (average daily amount for a normal adult)

Mango

EXCELLENT SOURCE OF VITAMINS A AND C
VERY LOW IN SODIUM
SOURCE OF FIBRE

↑ Brazilian mango

Origin

• Mangos have been cultivated in Asia for nearly 6,000 years. Originally from tropical areas of Asia, mangos, of the Anacardiaceæ family, were introduced by Arab traders into Africa in the 10th century. Portuguese explorers brought them to South America in the 16th century and they gradually spread throughout the tropical regions of the world. Indians have long believed that mangos have medicinal properties that aid digestion and purify the blood and skin. Most of our mangos come from Mexico, Central and South America, the Caribbean and Florida. Mangos are an important fruit in tropical countries. There are an estimated 14 million tons of them grown every year! In India alone there are about 1,400 varieties.

Varieties and description

• There are dozens of varieties of mangos, ranging from about 0.22 to 1.36 kilos (1/2 to 3 pounds) in weight. (Smaller sizes are almost never imported.) They are normally oval or kidney-shaped, and have a fibrous bright yellow to orange flesh that contains a large flat pit. The thin, leathery skin of ripe mangos ranges from green to yellow to orange to red, and variations of these colours. Some of the more popular varieties are:

• **Tommy Atkins;** oval, about 0.45 kilos (1 pound), yellow-orange skin with brilliant red blush.

• **Van Dyke;** oval, about 0.28 kilos (10 ounces), yellow with a red-orange blush.

• **Haden;** kidney shaped, about 0.4 kilos (14 ounces), greenish-yellow with pale orange blush.

• **Keitt;** large and round, 0.68 to 1.36 kilos (1 1/2 to 3 pounds), ranges from green to reddish-green to yellowish green when ripe, very juicy, nearly fibreless and tender to the touch.

• **Kent;** oval, about 0.64 kilos (1.4 pounds), yellow to apricot with a red blush, very juicy and fibreless.

• **Palmer;** oval, about 0.64 kilos (1.4 pounds), yellow with a rose blush.

• **Francis;** kidney shaped, about 0.28 kilos (10 ounces), yellow with a peach blush.

↓ Haitian mango

Availability

• May to August is the most plentiful season for mangos and Tommy Atkins and Haden are the most popular varieties. Haitian mangos are available all year; the season in South America runs from December to February; Mexican mangos can be found from April to September; and Florida mangos are most plentiful from June to September.

Selection

• Mangos are usually sold firm and will need time to soften. Look for clear, unblemished, taut skins and an aromatic scent.

Storage

• Leave at room temperature for a few days, until it yields to gentle pressure and gives off a heady, sweet, tropical aroma. The colour of the skin is not always a sign of ripeness, and speckled fruit indicates advanced ripeness, not spoiling. Only refrigerate when ripe, and they will last for a week or more. They are generally tastiest when eaten as soon as ripened. Mangos can be frozen only if prepared as ice cream or sorbet, or if cooked first in a syrup or sauce.

Preparation

• Eating mangos can be messy, until you gain experience. The Mexicans have a special mango fork, which resembles a letter opener. They use it to pierce the pit lengthwise, then they peel the mango and eat it like a lollipop. Be very careful if you try this with a normal knife. A more usual method is to cut off each half following the flat, oval pit with a sharp knife, then cut these halves in wedges that are easily peeled, then cut away any flesh remaining on the pit. Some people score each half with criss-cross slices, invert the piece, then slice the flesh off close to the skin. This gives you nice cube-shaped pieces for a fruit salad.

Suggestions

• Mango is great in fruit salad, as well as in chicken or turkey salad.

• Slices of mango compliment pork chops, ham steaks, duck and other poultry dishes.

• Mango is an excellent topping for cheesecake, shortcake, trifle, crêpes, granola, ice cream or cottage cheese.

• Under-ripe mangos can be grated or cubed and eaten as a salad, with salt and pepper, or they can be pickled.

• For an exotic and elegant salad, peel and dice 2 or 3 mangos and wash and slice a basket of strawberries. Dress with a little lemon juice, salad oil, and salt and pepper to taste. Serve on a bed of mixed red and green lettuce. Garnish with a few sprigs of finely minced fresh mint.

• Try growing a house plant from your mango pit.

• Have a mango-yogurt shake for breakfast; simply put the flesh of a mango, a little orange juice, some plain yogurt, and a spoon of honey in your blender, and mix well. Top with whipped cream for a special occasion. (See photo.)

Nutritional Value
Mango
(per 207 g = 7 1/4 oz)

			RDA**	%
calories*	(kcal)	135		
protein*	(g)	1.1		
carbohydrates*	(g)	35.2		
fat*	(g)	0.6		
fibre	(g)	1.7	30	5.7
vitamin A	(IU)	8060	3300	244
vitamin C	(mg)	57	60	95
thiamin	(mg)	0.12	0.8	15
riboflavin	(mg)	0.12	1.0	12
niacin	(mg)	1.2	14.4	8
sodium	(mg)	4	2500	0.2
potassium	(mg)	322	5000	6.4
calcium	(mg)	21	900	2.3
phosphorus	(mg)	22	900	2.4
magnesium	(mg)	18	250	7.2
iron	(mg)	0.26	14	1.8
water	(g)	169.1		

* RDA variable according to individual needs
** RDA: recommended dietary allowance (average daily amount for a normal adult)

Mango and chicken stir-fry

(serves 4)

Ingredients		
2	chicken breasts, skinned and boned	
2	ripe mangos, peeled and sliced	
60 mL	sunflower oil	1/4 cup
15 mL	soy sauce	1 TBSP
60 mL	pineapple juice	4 TBSP
15 mL	vinegar	1 TBSP
30 mL	brown sugar	2 TBSP
15 mL	cornstarch	1 TBSP
2 mL	ginger	1/4 tsp
1	green pepper, cut in strips	
1	red pepper, cut in strips	
125 mL	slivered almonds	1/2 cup
750 mL	hot cooked brown rice	3 cups

Instructions

- In a small bowl, combine soy sauce, pineapple juice, vinegar, sugar, cornstarch and ginger. Mix well.
- Cut chicken into medium cubes, then brown in hot oil, preferably in a wok.
- Add strips of green and red peppers and almonds to the wok and stir-fry until slightly tender.
- Add soy sauce mixture to the wok, and stir until sauce thickens.
- Reduce heat, add sliced mangos, stir, then remove from heat.
- Serve on a bed of hot brown rice.

South American mandarins

(Satsuma/ Dancy/ Malaquina/ Ellendale/ Honey tangerine "Murcott")

↑ Malaquina

EXCELLENT SOURCE OF VITAMIN C
LOW IN CALORIES AND FIBRE
CONTAIN IRON AND CALCIUM

Origin

• Though all citrus fruit probably originated in the tropical areas of China and Southeast Asia from a dozen or fewer indigenous varieties, they have been cross bred and hybridized again and again through the centuries until all specific origins have been obscured. Sailors helped spread seeds to other countries, and now most temperate zones produce citrus. Although a good deal of the citrus we consume is grown in the Unites States, the enormous increase of production in South America has extended the availability of these popular fruits.

Varieties and description

• The general category of mandarin citrus includes satsumas, tangerines, clementines, tangelos and mandarin oranges. Mandarins are closely related to oranges, but have a thinner, more easily peeled skin, few or no pips, and generally, a much sweeter flesh. They are often flattened at both ends. Any green discoloration of the skin has no bearing on the quality of the fruit, but is from a lack of cold nights while the fruit is ripening, as most of this type of fruit is grown in the tropics.

• **Satsumas** are a hybridized mandarin, originally developed in Japan, in the province of Satsuma. Their skin and flesh are light orange, and they rarely have pips. Most countries grow different varieties of satsumas, so don't be surprized if they don't all taste the same. Peru and Argentina have become major producers of this refreshing citrus fruit for our market, in addition to the United States, Spain and Morocco, among others.

• **Dancy** mandarins, which are grown in Argentina and the United States, have a thin, attractive, vermilion orange skin that is very easily peeled. It has very sweet orange flesh that contains only two or three pips.

• The **Malaquina** is a cross between a mandarin and an orange developed in Uruguay. About 10 centimetres (4 inches) in diametre, it is slightly flattened top and bottom. Its orange skin is sometimes tinged with brown, and is very easy to peel. Its flesh is pale orange, very sweet, and it contains few or no pips.

• **Ellendales** are another variety of mandarin that grow in Argentina and in Australia. They resemble Malaquinas in size and shape, but have a darker orange skin. They appeal to people who enjoy a slightly sharp citrus flavour.

• Honey tangerines or **Murcotts** were developed about 30 years ago by crossing a tangerine and an orange. Their skin and flesh are a deep rich orange colour. This sweet, juicy fruit usually contains quite a few pips. Brazil and Argentina are important producers of this delicious fruit, in addition to the United States.

Availability

• The South American satsuma season is from the end of April through June. Spanish and Moroccan satsumas are often imported in September and October, and American satsumas are available from November to January.

• Dancy tangerines come from Florida in December or January and the season lasts about a month; the Argentinian season, which also lasts a month, usually begins at the end of June or beginning of July.

• Malaquinas are available from July to September.

• Ellendales begin toward the end of July and last until the end of December.

• Murcotts, or honey tangerines, come from South America from late summer through October, and from Florida in January and February.

Selection

• Choose firm, heavy fruit. All citrus fruit is picked ripe and ready to eat.

Storage

• Store mandarins in a plastic bag in the refrigerator for a week or two, or keep at room temperature for about a week. As with all produce, they taste best when they are fresh.

Preparation

• Mandarins are easy to peel and easy to eat.

Suggestions

- Juice made from honey tangerines is terrific.
- Mandarins are an ideal fruit for lunch boxes and picnics.
- For a simple, yet sophisticated dessert, serve mandarin segments over vanilla or coffee ice cream, and add a splash of Grand Marnier.
- For a light, tasty and attractive salad, peel, segment and seed 4 mandarins. Coarsely chop 2 endives, dice 1 avocado, mince a few scallions and tear up a small lettuce. Toss together with 30 mL (2 TBSP) each olive oil and lemon juice, 15 mL (1 TBSP) each fresh grated ginger and honey, and salt and pepper to taste.
- Chopped mandarins are also terrific in rice salad. Save leftover cooked rice and mix in several peeled, seeded and chopped mandarins, along with 1 medium diced sweet onion, 1 or 2 minced celery stalks and 1 bunch of chopped watercress. Mix the following dressing and toss with the rice and vegetables: 45 mL (3 TBSP) each sunflower seed oil and white wine vinegar, 10 mL (2 tsp) Dijon mustard, and salt, pepper and celery salt to taste.
- Add mandarin segments to stir-fried beef or chicken dishes.
- Mandarin segments are very tasty in turkey or chicken stuffing.
- Mandarins are delicious in custard tarts, crêpe fillings, trifle and as decoration for cakes.

Nutritional Value
Mandarin
(per 1 medium = 85 g = 3 oz)

			RDA**	%
calories*	(kcal)	37		
protein*	(g)	0.5		
carbohydrates*	(g)	9.4		
fat*	(g)	0.2		
fibre	(g)	0.3	30	1
vitamin A	(IU)	773	3300	23.4
vitamin C	(mg)	26	60	43
thiamin	(mg)	0.09	0.8	11.3
riboflavin	(mg)	0.02	1.0	2
niacin	(mg)	0.1	14.4	0.7
sodium	(mg)	1	2500	0.04
potassium	(mg)	132	5000	2.6
calcium	(mg)	12	900	1.3
phosphorus	(mg)	8	900	0.9
magnesium	(mg)	10	250	4
iron	(mg)	0.9	14	6.4
water	(g)	73.6		

* RDA variable according to individual needs
** RDA: recommended dietary allowance (average daily amount for a normal adult)

Honey tangerine (Murcott) ↑

↓ Ellendale

Mandarin mousse

(serves 6)

Ingredients

175 g	sweet cooking chocolate	6 oz	15 mL	unflavoured gelatin	1 TBSP
4	large eggs, separated		45 mL	fresh mandarin juice	3 TBSP
30 mL	powdered sugar	2 TBSP	160 mL	35 % cream	2/3 cup
	zest from 2 mandarins		4-6	mandarins	

Instructions

- Attach wax paper around a soufflé or a deep serving dish so that it sticks up above the top by about 6 centimetres (2 1/2 inches).
- Break the chocolate into small pieces and melt it in a double boiler on low heat, stirring often.
- Let cool a little, then add the sugar and the beaten egg yolks, mixing well. (If the chocolate is too hot, the yolks might curdle.)
- Add the mandarin zest, and let cool.
- Mix the mandarin juice with the gelatin in a heavy skillet and place on low heat for 4 to 6 minutes without stirring.
- Gradually pour the mandarin-flavoured gelatin into the chocolate mixture, stirring constantly.
- Whip the cream, and beat the egg whites until stiff.
- Fold the cream and egg whites alternately into the other ingredients, a small amount of each at a time, just as the gelatin is beginning to set.
- Carefully transfer everything to your serving dish and refrigerate at least an hour.
- Peel mandarins and remove the membrane from each segment.
- Garnish the mousse with a few mandarin segments and perhaps a bit of grated chocolate. Serve each portion with fresh mandarin segments to add colour and fresh flavour.
- For a fancier, though more complicated dessert, make half the mixture with white chocolate and half with dark chocolate, and alternate layers, refrigerating 15 to 20 minutes between each stage.

Rainier and Montmorency cherries

↑ Rainier cherries

➠ Origin

• Originating in Asia thousands of years ago, the cherry, a member of the Rosaceæ (rose) family, is related to many fruits including the plum, apple, pear and peach. More than 600 distinct varieties have been identified, but only a few are cultivated for commercial purposes. A mere 35 % of the cherry harvest is consumed in its fresh state. Cherries are a favourite fruit of birds, who helped spread cherries to many temperate areas by carrying the seeds in their stomachs on long migrations. This delicious summer fruit was highly prized by the Romans, who introduced it throughout their empire. Early settlers brought the fruit to North America, where it is now grown on a very large scale.

➠ Varieties and description

• These two types of cherries couldn't be more opposite in taste and appearance.

• Rainier cherries are one of the most delicious and attractive varieties. This very large, sweet, juicy cherry has a yellow skin with a pink blush, and pinkish white flesh. California and Washington produce this delectable fruit.

VERY LOW IN CALORIES
LOW IN FIBRE
NO SODIUM
CONTAIN VITAMIN C

• Montmorency sour cherries, which are cultivated in Southern Ontario, are generally used in cooking. This excellent plump red cherry is sold fresh in small baskets, or washed, pitted, stemmed and packed in convenient plastic pails of 2.3, 5.5 or 11 kilos (5, 11 or 22 pounds) with 10 % sugar (by weight) to preserve their colour, flavour and texture.

➠ Availability

• Rainier cherries from California should appear about the end of May for three weeks; Washington's crop is usually ripe in June, and also has only a short three week season.

• Montmorency sour cherries run from mid-July to mid-August. The crop is limited and is sometimes pre-sold by advance orders.

➠ Selection

• Rainier cherries are very perishable because they are picked ripe, and bruise easily if not handled properly. Look for cherries with their stems still attached.

• The sour cherries are also very perishable and must be frozen. Generally, if a cherry looks good, it tastes great.

➠ Storage

• Rainier cherries are best enjoyed as soon as possible after purchase. Keep them refrigerated in a bowl, covered by a paper towel to absorb excess moisture. Although all sweet cherries can be frozen (after washing, pitting, stemming and sealing in bags), it is almost a shame to do this to such a superb fruit.

• Montmorency sour cherries must be frozen if not used immediately after purchase. Then you can enjoy them at leisure when the short season is over. Just divide your cherries into smaller packets, each one enough for a pie or dessert, and freeze. They'll taste fresh and delicious when defrosted.

Preparation

- Wash Rainier cherries just before you serve them.
- The Montmorency sour cherries that are sold in the basket must be pitted before you can use them in desserts. Most people prefer to bake with them.

Suggestions

- A bowl of fresh Rainier cherries is an exotic dessert in itself.
- Pit cherries, dip in chocolate sauce, and use them to garnish ice cream or cheese cake. (This is ideal for frozen cherries.)
- Try glazing a ham or joint of lamb with sour cherries before roasting.
- A sour cherry sauce is delicious with roast chicken, duck or turkey. To make enough sauce for 6 to 8, melt a little margarine or butter and sauté 1 finely chopped medium onion until soft but not browned. Add 500 mL (2 cups) pitted Montmorency sour cherries (drain and reserve liquid if cherries are frozen), 125 mL (1/2 cup) sugar and an equal amount of reserved liquid (or stock, if cherries are fresh), 60 mL (4 TBSP) each soy sauce and Chinese black or balsamic vinegar, a little grated fresh ginger and citrus zest, 1 crumbled chicken stock cube, 2 cloves, and a good dash of grated nutmeg and salt. Bring to the boiling point, then simmer for a few minutes. In a cup, mix 30 mL (2 TBSP) cornstarch with enough reserved liquid to make a smooth paste. Add the cornstarch to the cherry sauce and stir until the sauce thickens. Serve sauce over poultry or pork.

Nutrient Value
Rainier cherries
(per 68 g = 2 1/2 oz)

			RDA**	%
calories*	(kcal)	49		
protein*	(g)	0.8		
carbohydrates*	(g)	11.3		
fat*	(g)	0.7		
fibre	(g)	0.3	30	1.0
vitamin A	(IU)	146	3300	4.4
vitamin C	(mg)	5	60	8.3
thiamin	(mg)	0.03	0.8	3.8
riboflavin	(mg)	0.04	1.0	4
niacin	(mg)	0.3	14.4	2.1
sodium	(mg)	0	2500	0
potassium	(mg)	152	5000	3
calcium	(mg)	10	900	1.1
phosphorus	(mg)	13	900	1.4
magnesium	(mg)	8	250	3.2
iron	(mg)	0.26	14	1.9
water	(g)	54.9		

*RDA variable according to individual needs
**RDA: recommended dietary allowance (average daily amount for a normal adult)

Deep dish sour cherry pie

(serves 8)

The pastry for this pie has to be chilled for at least an hour before you use it. It is made with sour cream instead of water, which makes it a little harder to roll — but the great taste is worth it. You may find it easier to roll the lattice strips needed for the top of the pie by hand than with a rolling pin.

Ingredients		
For pastry		
375 mL	unbleached flour	1 1/2 cups
2 mL	salt	1/4 tsp
120 mL	margarine or butter	8 TBSP
45 mL	sour cream	3 TBSP
For filling		
1 kilo	sour cherries	2 1/2 lbs
250 mL	sugar	1 cup
45 mL	flour	3 TBSP
10 mL	lemon juice	2 tsp
5 mL	cinnamon	1 tsp
15 mL	milk	1 TBSP
15 mL	sugar	1 TBSP

Instructions

- To make the pastry for a lattice topping, mix the flour and salt in a bowl.
- Break the well chilled margarine or butter into small pieces. Add to the flour and, using 2 knives, rapidly cut into small pea-sized pieces. Then, using your fingers, lightly break it down to a finer consistency. The trick is to handle the mixture as little as possible, in order to ensure a nice texture for your crust.
- Add the sour cream little by little, only using enough to moisten the crust so that it will stick together.

• Roll the dough and cut strips for a lattice topping.

• Thaw and drain the frozen pitted cherries. Keep the juice for another recipe, or use to make a cherry drink, by mixing it with soda water.

• Mix the sugar, lemon juice, and cinnamon into the cherries, then gently sprinkle the flour over them, being careful not to make lumps. Mix well.

• Place cherries in a deep glass baking dish; top with latticed pastry strips.

• Glaze the pastry; using a pastry brush (or your finger), paint a light coating of milk on to the pastry, then sprinkle on a tablespoon of sugar.

• Bake in a preheated oven at 190 °C (375 °F) for about 40 minutes, lowering the temperature slightly about half way through.

Montmorency sour cherry jam

Ingredients

1 L	sour cherries	4 cups
750 mL	sugar	3 cups
1/2	lemon	

Instructions

• Cut the lemon in thin slices, then quarter each slice.

• Put the pitted cherries and the lemon slices in a large, heavy saucepan and cover with the sugar. Put aside and let stand until the sugar dissolves.

• Boil rapidly for about 15 minutes, stirring constantly, then simmer until the jam thickens. To test for the right consistency, drop a teaspoon of the jam onto a small plate and, as it cools, see if it holds its shape or is still too runny.

• After bottling and cooling, you can keep extra jars of jam in the refrigerator or in the freezer until you require them.

Lychee and longan

Longan ↑

EXCELLENT SOURCE OF VITAMIN C
LOW IN FIBRE AND SODIUM
TRACES OF CALCIUM, MAGNESIUM AND IRON

Origin

• These similar fruits are now popular in North America, thanks to the efforts of an American missionary who brought them back from China earlier this century.

• Originally from southern China, lychees, litchis, or Chinese cherries have been cultivated for over 2,000 years. Lychees and longans are both members of the tropical Sapindaceæ (saponin) family, along with other exotic fruits such as rambutans and akees. Lychees are still given as good luck gifts for Chinese New Year. Today they are grown for export in China, India, Malaysia, Taiwan, Thailand, the Philippines, Zimbabwe, South Africa, New Zealand, Australia, Israel, Mexico and Florida.

• Longans, lungans, or dragon's eyes, which are closely related to lychees, originated in India but are now most heavily cultivated in China. Longans are exported from Taiwan and other Asian countries, and are being grown on a small scale in Florida and California.

Varieties and description

• Lychees are small round fruits, usually 2.5 to 4 centimetres (1 to 1 1/2 inches) in diametre, and longans are even smaller. The lychee's bumpy pink or red shell is soft when freshly picked, but turns hard after a day or two, and gradually turns brown. Its pearly white, translucent flesh is refreshingly juicy, very sweet and extremely aromatic. Longans have a smooth, orange shell, that turns light yellowish-brown and hardens after harvesting. Both fruits have a brittle shell that is easily peeled and a large inedible seed. The flesh of these delicious exotic fruits is similar looking and tasting.

Availability

• As so many countries in different parts of the world grow them, lychees are now available just about all year long. Longans can be found mainly in August and September.

Selection

• Look for fruit that shows no sign of cracking, and no spots of its juice on the outer shell. They are sometimes sold on the branch, which is very attractive.

Storage

• Refrigerated in a plastic bag with a paper towel to absorb any moisture, these fruits should last a week or more, though their delicate perfume may diminish. If kept too long, they may begin to ferment and taste acidic. They are best eaten as fresh as possible.

Preparation

• Simply crack the shell at the stem end, peel and enjoy. If used in recipes, discard the seed first and add the fruit toward the end of preparation to preserve its unique taste.

Suggestions

• A bowl of fresh lychees or longans is a simple, yet perfect way to end a meal.

• These fruits combine well with others, such as grapes, kiwis, pineapple, banana and tangerines, for an exotic fruit salad.

• Add a few to a stir-fry.

↓ Lychee

Longans with liqueurs

(serves 4)

Ingredients		
500 mL	rose water	2 cups
160 mL	fine sugar	2/3 cup
45 mL	green Crème de menthe	3 TBSP
125 mL	Poire William or kirsh	1/2 cup
16	longans, shelled and pitted	
3	fresh edible roses	
	a few chocolate leaves	

Instructions

- Mix the sugar and rose water, bring to a slow boil and reduce, until it is a syrup, then let cool to room temperature.
- Add the liqueurs and the longans, and let macerate in the refrigerator for about 2 hours.
- Just before serving, garnish with rose petals and chocolate leaves.

Lobster navarin with lychees

(serves 2)

Ingredients		
2-3	lobsters, depending on size	
12	lychees, peeled, pitted and halved	
45 mL	butter	3 TBSP
60 mL	dry rosé wine	1/4 cup
45 mL	sauternes wine	3 TBSP
1	fresh mint leaf, finely chopped	
250 mL	35 % cream	1 cup
	salt and pepper to taste	

Instructions

- Buy precooked lobsters or cook live ones in boiling water, remove the flesh from the shell and reserve any attractive parts of the shell for decorating your platter.
- Melt the butter in a skillet, add the lobster flesh, and cook for about 3 minutes on high heat.

- Set aside the lobster, discard any remaining butter, add both types of wine to deglaze the pan and reduce at high heat for a few minutes.
- Add the mint and the cream and lower the temperature to medium, stirring continually. When the sauce begins to thicken, add the lychees, the lobster, salt and pepper, and serve immediately.

Nutritional Value

Lychee

(per 10 medium = 100 g = 3 1/2 oz)

			RDA**	%
calories*	(kcal)	66		
protein*	(g)	0.8		
carbohydrates*	(g)	16.5		
fat*	(g)	0.4		
fibre	(g)	0.2	30	0.7
vitamin A	(IU)	0	3300	0
vitamin C	(mg)	72	60	120
thiamin	(mg)	0.01	0.8	1.3
riboflavin	(mg)	0.07	1.0	7
niacin	(mg)	0.6	14.4	4.2
sodium	(mg)	1	2500	0.04
potassium	(mg)	171	5000	3.4
calcium	(mg)	5	900	0.6
phosphorus	(mg)	31	900	3.4
magnesium	(mg)	10	250	4
iron	(mg)	0.31	14	2.2
water	(g)	81.8		

*RDA variable according to individual needs
**RDA: recommended dietary allowance (average daily amount for a normal adult)

Fresh fig

MODERATE IN CALORIES
SOURCE OF THIAMIN AND RIBOFLAVIN

↑ Green fig

Origin

• The fig is an important member of the Moraceæ family, which also includes mulberries, jackfruit and breadfruit. Mentioned by an Egyptian pharmacopeia about 5,000 years ago, the fig originated in the southwestern region of Turkey, in an area that was once part of the ancient kingdom of Lydia. It was an important crop to the early civilizations in the Mediterranean area. The Greeks described it as a source of strength, and it is frequently mentioned in the Bible. The fig was a famous symbol to the Romans, as Romulus and Remus, the founders of Rome, were found sheltering under a fig tree by the wolf who raised them. The Romans later introduced the fig, which they called ficus, into all the temperate zones of their great empire. From Europe, the Spaniards introduced figs to their settlements in the New World, especially in St. Augustine, Florida, and later in California, and the French brought them to New Orleans. Today, we import figs mostly from California, Italy, Greece, Turkey and Brazil.

Varieties and description

• Figs are not, botanically speaking, true fruit, but rather fleshy receptacles for dozens of little edible seeds, which are the true fruit of the fig tree. A fresh fig is about the size of a medium plum, and is shaped like a teardrop. When ripe, they should be soft, sweet and delectable. The skin is edible. There are more than 150 varieties of figs, which come in all colours — white, yellowish green, green, red, violet, purple, brown and black. The most popular varieties are Black Mission, Kadota and Calmyrna. Figs are very fragile when fresh, and are usually flown to market.

Availability

• The three varieties mentioned above, which are grown in California, are at their peak between June and November. Fresh figs are available just about all year long from other countries, subject to demand.

Selection

- Choose plump figs of a uniform colour and handle them carefully.

Storage

- Keep refrigerated and use as soon as possible.

Preparation

- Wash and dry carefully, then twist or cut off the hard tip of the stem if you are going to cut the figs or use them in a recipe, or leave as is to serve them whole.

Suggestions

- Fresh figs are excellent served with prosciutto.
- Figs are delicious in a fruit salad, especially with melon.
- Cut figs in quarters and serve with cheese and crackers.
- Serve a bowl of figs and strawberries with a cheese soufflé.
- The French make a superb dessert by baking figs in white wine until they are soft, and serving them with fresh thick cream.
- The Italians sometimes make slits in their figs and stuff them with a mixture of soft cheese (such as Dolcelatte, Mascarpone, any creamy cheese, or even Stilton) mixed well with a little sweet wine.

Nutritional Value

Fresh fig

(per 2 medium = 100 g = 3½ oz)

			RDA**	%
calories*	(kcal)	74		
protein*	(g)	0.8		
carbohydrates*	(g)	19.2		
fat*	(g)	0.4		
fibre	(g)	1.2	30	4
vitamin A	(IU)	142	3300	4.3
vitamin C	(mg)	2	60	3.3
thiamin	(mg)	0.06	0.8	7.5
riboflavin	(mg)	0.06	1.0	6
niacin	(mg)	0.4	14.4	2.8
sodium	(mg)	2	2500	0.08
potassium	(mg)	232	5000	4.6
calcium	(mg)	36	900	4
phosphorus	(mg)	14	900	1.5
magnesium	(mg)	16	250	6.4
iron	(mg)	0.36	14	2.6
water	(g)	79.2		

*RDA variable according to individual needs

**RDA: recommended dietary allowance (average daily amount for a normal adult)

Chinese greens

Origin

• Bok choy, baby bok choy, bok choy sum, yow choy sum, gai lon, Shanghai bok choy, gai choy and a number of other green vegetables are all types of oriental cabbages and members of the large Cruciferæ family. For the most part, they look and taste different from each other, though getting all their names right is a difficult feat.

• One of the most popular of this group, bok choy is also known as Chinese chard, Chinese white cabbage, pak choi, paksoi and celery cabbage. It was once grown for its seeds, which contain an edible oil. Many varieties of bok choy and other Chinese cabbages have been cultivated for thousands of years in China and various East Asian countries, but they only became known in Europe at the beginning of the 18th century.

• Chinese immigrants to the United States brought bok choy to California and today it is grown in large quantities in New Jersey and in Florida. American agricultural scientists improved existing varieties of various Chinese cabbages not long after World War I, then reintroduced them to China.

• Bok choy sum or choy sum, a closely related vegetable of similar origins, is also known as flowering bok choy, or flowering white cabbage. It is considered more of a delicacy than bok choy.

• Yow choy sum, confusedly called choy sum by some people, is also known as flowering greens. Just to add to the confusion, in some parts of China they call bok choy "yow choy".

• Gai lon, another similar-looking vegetable, is sometimes known as Chinese kale, Chinese broccoli or gai lan.

• Mustard greens, such as gai choy, dai gai choy and gai choy sum, are related to all these vegetables. Their pungent flavour is excellent in soups and stir-fries.

Varieties and description

• At first glance, bok choy may remind you of a celery crossed with Swiss chard. It has dark green leaves growing from its succulent, central, thick white stalks. There are long and short-stemmed varieties, and short bulbous varieties. The stalks have a crispy texture and a mild, sweet, refreshing taste, while the leaves are similar to Swiss chard and mild cabbage. Baby bok choy, a very popular variety, can range in size from small to large. It is sometimes cooked whole or in halves.

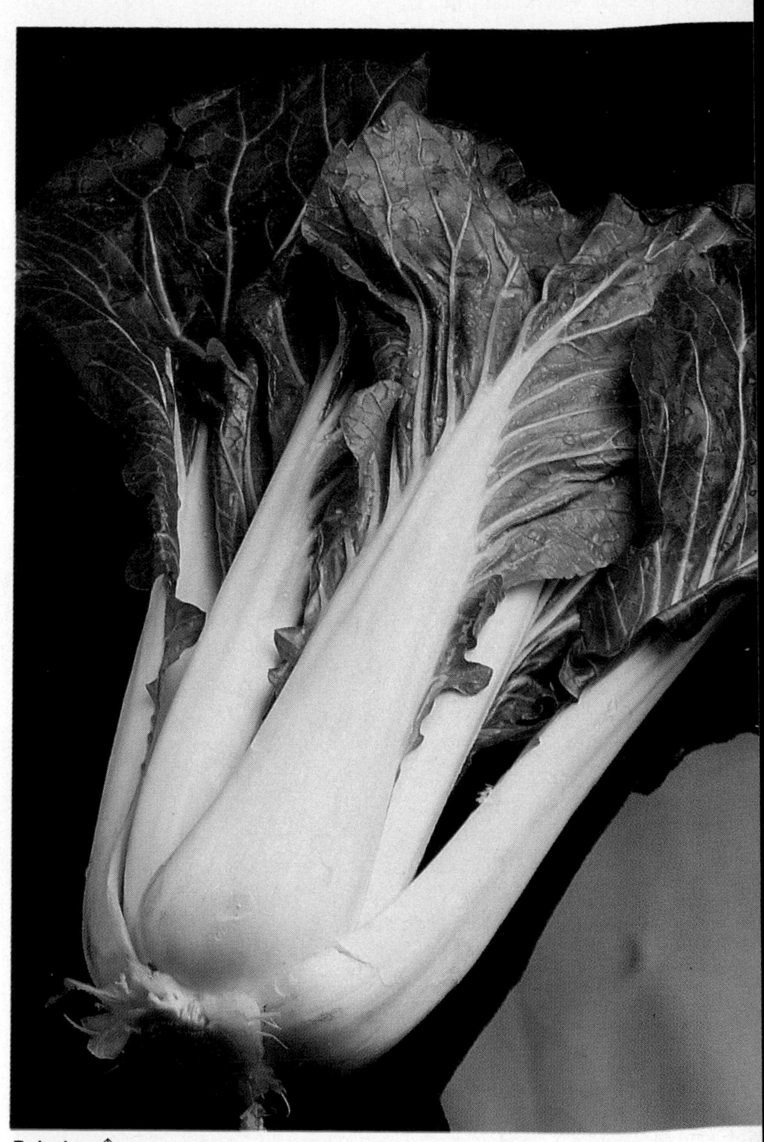

Bok choy ↑

• Bok choy sum looks very much like bok choy, except that it is smaller and has little yellow flowers at the end of some of its stalks; these are cooked and eaten along with the rest of the vegetable. Compared to bok choy, the stalks are much narrower and the leaves not as dark green.

• Yow choy sum, commonly abbreviated to yow choy, is probably the vegetable you get when you order Chinese greens in a restaurant. Yow choy has even smaller yellow or purple flowers, which are also edible, and long, narrow stalks topped by long, oval-shaped, dark green leaves that have prominent veins. It is agreeably bitter and very palatable.

• Gai lon is a sort of flowering broccoli, with very small, white flowers, very large dark green leaves and slender stalks. The entire vegetable is eaten and has a pleasant flavour.

Availability

• These vegetables are available year round. They are not always plentiful because the demand for them is still not great outside the Asian community.

↓ Yow choy sum

Selection

• Buy bok choy with fresh-looking leaves and firm bright white stalks. All these vegetables should have firm stalks and none should have limp leaves.

Storage

• Use as fresh as possible, not washing until just before use. You can keep them for a day or two in plastic bags in the refrigerator.

Preparation

• Discard a slice from the base of the bok choy and separate the stalks. Cut the leafy part off each stalk. Cut the stems on the diagonal in large pieces and cut or tear the leaves into a more manageable size.

• You can use both parts in the same recipe, or you can save one part (unwashed) for another meal.

• Cooking time should only be a matter of minutes or you will spoil its wonderful texture. The leaves cook so quickly that they should not be added until the very end of preparation.

• Treat bok choy sum and yow choy sum in the same way. Prepare gai lon as you would prepare broccoli or stir-fry it.

Suggestions

• Steam, boil or sauté and add a squeeze of lemon, a little butter and salt, or Chinese vinegar and a few drops of sesame oil.

• All these vegetables are excellent in a simple stir-fry, alone or with other oriental vegetables. Season with a little garlic, soy sauce, and perhaps ginger, and serve with noodles or rice. Add slivers of beef, chicken or shrimps for a main course.

• The stems of bok choy are especially good cooked in a clear soup with a touch of soya sauce; shred the leaves finely and add to the soup at the end.

• Steam or boil coarsely chopped bok choy (or another Chinese green) for 2 or 3 minutes, drain immediately, toss with the following and serve: Mix together 5 mL (1 tsp) each water and dry mustard. Add 5 mL (1 tsp) fresh grated ginger, 30 mL (2 TBSP) soy sauce and a pinch of sugar and salt.

• Bok choy is very tasty blanched and baked au gratin.

• Gai lon is outstanding used in both oriental and western dishes.

Gai lon in oyster sauce

(serves 4)

Ingredients		
2	bunches of gai lon	
4-8	water chestnuts	
12	mushrooms	
4	scallions (green onions)	
30 mL	peanut oil	2 TBSP
45 mL	oyster sauce	3 TBSP
15 mL	rice wine	1 TBSP
	pinch of sugar	
	pinch of salt	
	handful of toasted pine nuts	

Gai lon ↓

Instructions

- Chop gai lon in lengths of about 5 centimetres (2 inches). Slice water chestnuts and mushrooms. Mince scallions.
- Blanch or steam the gai lon for just a few minutes.
- In a wok, heat oil and add mushrooms and scallions; stir for a minute.
- Add gai lon and water chestnuts; stir for a minute
- Add oyster sauce, rice wine (sake), sugar and salt; stir for at least a minute. Keep stirring until done, which won't take long.
- Sprinkle with pine nuts or substitute toasted almond slivers.
- Serve as a vegetable side dish with rice and a main course.

Snow pea

EXCELLENT SOURCE OF VITAMIN C
GOOD SOURCE OF FIBER AND IRON
LOW IN CALORIES AND SODIUM

Origin

• From the Mediterranean region to the Middle East, the pea has been an important food staple for at least 5,000 years — a long history that the delectable snow pea does not share. Edible pea pods, members of the Legume family, were probably developed by the Dutch about 400 years ago. The snow pea, also known as the Chinese pea or mange-tout, was ideally suited to growing conditions in Asia and quickly became a very popular vegetable in China and other East Asian countries.

• California as well as Guatemala and other Central American countries have become major suppliers. Canada produces some snow peas for local markets.

Varieties and description

• Snow peas are particularly valued for their sweet, crisp, edible pods, rather than for the tiny peas they contain. About the same size as regular pea pods, they are very flat; the outline of the peas within should hardly be noticeable.

Availability

• They can be found year round, but the peak season is from May to September.

Selection

• Choose thin, flat, fresh-looking, bright green pods. The pods should be flexible to the touch, but crisp in texture. Excessive moisture or wrinkling should be avoided.

Storage

• Snow peas are highly perishable and should always be bought for immediate use. Keep them unwashed in a plastic bag in the refrigerator for a day or two.

Preparation

• Wash, then snap off a tiny piece from each end of the pod. Snow peas are delicious steamed or lightly blanched, but beware of cooking them for longer than a couple of minutes or you will destroy their bright green colour and lose their characteristic crispy texture.

Suggestions

• Snow peas are a quick and nutritious vegetable side dish when simply steamed, blanched or boiled a minute or two, and served plain, or with a dab of butter and a sprinkle of salt.

• Snow peas are excellent in a large variety of oriental dishes.

• Try them blanched in salads, with an appetizing vinaigrette.

• For a beautiful and delicious platter of hors-d'œuvres, wrap blanched snow peas around cooked shrimp, skewer with toothpicks and provide a tasty dip.

• Snow peas combine well with Chinese mushrooms and bamboo shoots.

• As a tasty party snack, slit open one side of each snow pea, blanch them, and stuff with a savory mixture such as cream cheese and anchovies.

Snow pea stir-fry

(serves 4)

Ingredients		
2	chicken breasts, deboned	
	juice of 1 lemon	
30 mL	soy sauce	2 TBSP
15 mL	fresh ginger, grated	1 TBSP
2	garlic cloves	
250 g	snow peas	1/2 lb
	half a bok choy	
	half a lo bok (daikon)	
5	stalks Chinese celery	
1	onion	
1	red pepper	
30 mL	peanut oil	2 TBSP
250 mL	chicken stock	1 cup
30 mL	water	2 TBSP
10 mL	cornstarch	2 tsp
	salt and pepper to taste	

Instructions

• Cut the chicken into strips. Combine the lemon juice, soy sauce, ginger and garlic. Marinate the chicken in the mixture for at least an hour.

• Wash and drain all the vegetables. Chop the bok choy coarsely, the daikon in thin rounds, the celery in short sticks, the onion finely and the pepper in slivers.

• Remove the chicken from the marinade, reserving the liquid.

• In a wok, heat the oil and brown the chicken quickly, just until it looses its pink colour. Add the onions and celery; stir a minute. Add the snow peas, red pepper and daikon; stir a minute or two. Add the bok choy, the chicken stock and the marinade. Let simmer about 2 minutes, taking care that the vegetables remain crisp.

• Mix the water and cornstarch. Add it to the wok and stir until the sauce begins to thicken, then remove the wok from the stove. Season if necessary.

• Serve with white or brown rice.

Nutritional Value
Snow pea
(per 100 g = 3 1/2 oz)

			RDA**	%
calories*	(kcal)	53		
protein*	(g)	3.5		
carbohydrates*	(g)	12.1		
fat*	(g)	0.2		
fibre	(g)	1.9	30	6.3
vitamin A	(IU)	45.6	3300	1.4
vitamin C	(mg)	60	60	100
thiamin	(mg)	0.15	0.8	18.7
riboflavin	(mg)	0.08	1.0	8
niacin	(mg)	1.05	14.4	7.3
sodium	(mg)	3.9	2500	0.16
potassium	(mg)	200	5000	4
calcium	(mg)	43	900	4.7
phosphorus	(mg)	0	900	0
magnesium	(mg)	0	250	0
iron	(mg)	2	14	14
water	(g)	89		

*RDA variable according to individual needs
**RDA recommended dietary allowance (average daily amount for a normal adult)

Fresh ginger

Origin

• The true origins of ginger are unknown, although most botanists believe it comes from the tropical zones of Southeast Asia. We do know it has been cultivated in Southern China and in India for thousands of years.

• In the 13th century, Arab traders and explorers brought roots from Asia to start ginger cultivation in East Africa, where they had extensive settlements. About three centuries later, Portuguese traders introduced ginger to West Africa and began a lucrative trade with other European nations. Ginger became extremely popular, especially in England. Spanish explorers later brought it from Europe to the New World.

• Before refrigeration, ginger was considered an essential food preservative. In Asia, ginger has long been esteemed for its medicinal properties.

• Today, Hawaii, Fiji, Brazil, Puerto Rico, Florida, Jamaica and New Zealand are some of the largest producers of fresh ginger for the North American market.

Varieties and description

• Although we often call it ginger root, ginger is actually a rhizome, or underground stem, which has true roots and shoots that are removed after harvesting. A "hand" of ginger is very irregularly shaped, with little knobs growing out of each other. Its skin is light brown and quite thin and its flesh is beige, with a characteristic spicy flavour.

• Depending on the mineral content of the soil in which it has grown, ginger can be tinged with blue, green or gold. It is closely related to the spices tumeric and galangal.

• Young ginger is tender and very pretty, with pale pink and green shoots. It's so mild that you can chew on a piece of it — something you wouldn't dream of doing with mature ginger.

Availability

• Ginger can be found year round.

Selection

• Mature ginger should be firm and heavy when fresh, with smooth skin. Look for a full hand, with the least number of protruding knobs. Young ginger should have delicately coloured shoots and translucent skin.

Storage

• There are many ways to store fresh ginger, depending on how long you intend to keep it. It can be stored uncovered in the refrigerator for a week or two, but then it may begin to dry out. Some people freeze it and break off chunks as needed; others put it in a jar cut in small pieces, cover it with sherry and keep it refrigerated. It also keeps fresh when buried in a small pot of sand.

Preparation

• Peeling ginger is optional. Depending on your recipe, slice it finely, mince it or grate it. Substitute about 15 mL (1 tablespoon) fresh ginger for 1 mL (1/8 teaspoon) dried ginger.

Suggestions

• Puréed carrot and ginger soup tastes wonderful and looks beautiful. Simply sauté 2 large chopped onions,

1 kilo (2 pounds) chopped carrots, 30 to 45 mL (2 to 3 tablespoons) grated ginger, and 2 cloves of garlic in a little butter or oil. Add 1 litre (4 cups) of chicken or vegetable stock, and purée everything when the carrots are soft. Heat, add salt and black pepper to taste; serve, garnishing with chopped parsley or chives.

- Gingerbread, ginger cake, candied ginger, ginger snaps, ginger puddings, pies and muffins are popular desserts. Crystallized ginger dipped in dark chocolate is excellent.

- Ginger accompanies beef, chicken, seafood and vegetable dishes equally well.

- It's an essential ingredient for oriental and Indian cuisine. Pickled ginger is one of the most widely used Japanese condiments.

- Delicious in barbecue sauces, chutney, curry and spicy marinades, a little grated fresh ginger is also very good added to salad dressings. To make an oriental dressing, excellent with blanched snow peas and nappa, add some grated ginger to 30 mL (2 TBSP) light soy sauce, 1 crushed garlic clove, 15 mL (1 TBSP) each vegetable oil, sesame oil, Chinese black or balsamic vinegar and salt and pepper to taste.

- A hot ginger and honey drink is a great home remedy for a sore throat and ginger beer, ginger ale and ginger wine have long been popular drinks.

↓ **Young ginger**

Nappa (Chinese cabbage)

VERY LOW IN CALORIES
SOURCE OF VITAMIN C
TRACE OF CALCIUM AND IRON

Origin

- Also called celery cabbage, pet-sai, tientsin, sui choy, chow choy and won bok, nappa has been a staple food in a number of eastern countries for many centuries. Texts that are approximately 3,000 years old refer to cabbage as a cure for male baldness. Nutritionists believe that members of the cabbage family (Cruciferæ) are essential to a healthy diet. Throughout Asia, pickled cabbage is a regular part of the normal diet.

- Today, with the growing popularity of this sweet, mild cabbage, nappa is being cultivated for our market mainly in California, Florida and New Jersey.

Varieties and description

- Nappa is a long-leafed cabbage, more like Romaine lettuce in shape than like our round cabbages. It has a much lighter flavour than head cabbage, and is juicier and more tender. Pale green to white in colour, nappa leaves have crinkled edges with prominent veins growing from wide stalks or ribs.

- There are two main varieties. One is longer and more slender and its leaves curl outwards; the other is shorter and stockier, with leaves curling inward.

- If you don't like cabbage, you should still try nappa, which has a much more pleasing flavour.

- Nappa grows up to about 25 centimetres (10 inches) long, and shahali (or Chinese lettuce, which looks very similar to nappa) grows longer.

- As there are no other discernible differences to the ordinary palate, most people call both nappa.

Availability

- Nappa can be found year round, but is most plentiful from October to December. May is the lowest point in its season.

Selection

- Choose fresh-looking leaves that seem firm.

Storage

- Store nappa in a plastic bag in the refrigerator and it will keep well for about a week or two.

- For use in salads, it is best to eat it within a few days of purchase.

- Do not wash until ready to use.

Preparation

- Discard a small slice from the base of the nappa and remove as many leaves as you need. Wash, drain, then use cooked or raw. It can be cut in a variety of ways, depending on your recipe. Nappa never gives off strong odours when cooking.

- Nappa is commonly used to make a spicy side salad that accompanies many meals in the Orient. Simply chop a nappa coarsely, sprinkle with salt, and let stand a few hours, turning occasionally, until it loses its crisp texture. Squeeze out as much water as you can. Mix in a handful of minced scallions, 2 to 3 cloves of crushed garlic, a knob of grated ginger, a dash of chili, rice vinegar, soy sauce, and a little sugar and salt. After marinating about a day, refrigerate until used. Sprinkle each serving with a few drops of sesame oil.

Suggestions

- Shred and add nappa to salad or cole slaw.

- Slice and add to a stir-fry or a stew.

- It is delicious in a sweet and sour sauce. In a large pot, mix chopped nappa, diced tomatoes, sweet onions and a grated carrot or two, with minced garlic, grated ginger and a little oil, sugar and vinegar. Add a touch of chili, if you like spicy food. Cook everything together about 15 minutes, stirring often. A few small pieces of pineapple will add a good flavour.

- Shred and add nappa to a clear broth or an oriental soup toward the end of cooking.

- To make a simple, tasty soup, chop up a small nappa and lightly sauté it with diced onion, minced garlic and finely chopped fresh coriander. Add water and stock cubes, and season with a dash of oyster sauce, rice wine, salt and pepper.

- Chop up nappa and steam it in about 1½ to 2 centimetres (1 inch) water for about 5 minutes. Serve with a little butter and salt as a delicious vegetable side dish.

Nutritional Value
Nappa
(per 100 g = 3½ oz = approx. 2 cups)

			RDA**	%
calories*	(kcal)	14		
protein*	(g)	1.2		
carbohydrates*	(g)	3.0		
fat*	(g)	0.1		
fibre	(g)	0.6	30	2
vitamin A	(IU)	150	3300	4.5
vitamin C	(mg)	25	60	42
thiamin	(mg)	0.05	0.8	6.3
riboflavin	(mg)	0.04	1.0	4
niacin	(mg)	0.6	14.4	4.2
sodium	(mg)	23	2500	0.9
potassium	(mg)	253	5000	5.1
calcium	(mg)	43	900	4.7
phosphorus	(mg)	40	900	4.4
magnesium	(mg)	14	250	5.6
iron	(mg)	0.6	14	4.3
water	(g)	95		

*RDA variable according to individual needs
**RDA recommended dietary allowance (average daily amount for a normal adult)

Oriental radish

RICH IN VITAMIN C
VERY LOW IN CALORIES AND SODIUM
LOW FIBRE CONTENT

Origin

• These radishes are known as daikon to the Japanese and as lo bok, lo pak, Chinese radish or Chinese turnip to the Chinese.

• One of the fastest growing vegetables in the world, the radish has been known to mankind since prehistoric times.

• All types of radishes are members of the Cruciferæ family and probably originated in the Far East — possibly China. Oriental radishes are exceptionally popular throughout Asia. They are now grown in many parts of the world, including North America.

Varieties and description

• Daikon is an extremely long (usually more than 30 centimetres or 1 foot), smooth, white, carrot-shaped root. The skin is sometimes black, pink or even green, although these coloured varieties are rarely seen in North America. Inside its very thin skin, the flesh is white and very juicy, with a mild, refreshing, radishy taste and a crunchy texture. Daikon leaves and sprouts are very popular in Japan but hard to find here.

Availability

• It can be found all year long, but daikon available during the winter months is considered tastier.

Selection

• Choose firm, smooth-skinned roots.

Storage

• Keep daikon wrapped in a plastic bag in the refrigerator for 3 or 4 days until it begins to lose its juiciness. You don't have to use it all at once. If kept longer, it should be used cooked instead of raw.

Preparation

• Scrub the daikon or pare a thin layer off the part you are using. The Japanese grate it and serve it as a salad with most meals, especially with fried foods like tempura, as they believe it cuts down the oiliness of fried foods and aids digestion. It is often pickled, usually eaten raw and sometimes cooked. Take care not to overcook it.

Suggestions

• Cut daikon in rounds or sticks and serve it with a dip.

• Slice or dice it and add it to salads.

• It can be added to a stir-fry or a stew, but it will lose its characteristic flavour and taste more like turnip.

• In soups, it can be lightly cooked and puréed, minced or julienned and added at the end.

• Try shredding daikon and adding it to your usual hamburger mixture, for a tastier and juicier burger.

• For a delicious salad, try chunks of daikon along with your favourite raw vegetables, seasoned with yogurt, lemon juice, garlic, salt and pepper; for an extra zing, add fresh watercress.

• Daikon is excellent made into Japanese pickles: slice a daikon very thinly, mince a small onion, add a pinch of chili, the juice of half a lemon, about 60 mL (4 tablespoons) rice vinegar and a pinch of sugar, salt and pepper. Mix everything thoroughly, and let stand for two days in a cool place. Keep it in the refrigerator for about two weeks.

Oriental mushrooms

↑ Oyster mushrooms

GOOD SOURCE OF RIBOFLAVIN
SOURCE OF PHOSPHORUS, IRON AND POTASSIUM
VERY LOW IN CALORIES

Origin

• Mushrooms existed millions of years before mankind. They grow from minuscule spores, rather than from seeds, which are easily transported enormous distances by the wind. It is not always possible to say where they originated, as the same varieties exist in many parts of the world.

• Though cultivation of mushrooms in the West didn't begin until the 18th century, they have been gathered from the wild since ancient times.

• Out of the approximately 2,000 varieties of edible mushrooms, only a very few are available commercially.

• In recent years, there has been a growing demand in the West for popular varieties of oriental mushrooms such as oyster, shiitake, enoki and wood ear.

• Another sought-after oriental variety, the pine mushroom, is rarely available, although attempts have been made to grow it in the Pacific Northwest.

• Shiitake mushrooms have been cultivated for at least 2,000 years in Japan and were once available only to the powerful and wealthy Japanese elite.

• Oyster mushrooms grow in many parts of the world and are much sought after by Italians, Hungarians and other Europeans.

• The Chinese considered wood ear mushrooms excellent for the circulation. Wood ear mushrooms used to be available only in dried form from China. Though they are now grown in California, fresh ones are still hard to find.

• Oyster and shiitake mushrooms are now cultivated in Canada and the U.S. Enoki and wood ear mushrooms are both grown in the U.S.

Varieties and description

• Oyster mushrooms — also called pleurote, abalone mushrooms, tree oysters and shimeji — grow in clumps on the bark of older trees. They have very short stems and are often shaped like fans, growing one on top of another. They are beige or range from light to dark gray in colour and have very pronounced gills on the underside. Oyster mushrooms are usually fairly large.

• Shiitake mushrooms, also called golden oak or Chinese black mushrooms, used to be grown on a certain variety of oak tree. Until recently only dried ones were available in the west. Today, these brown-to-black, umbrella-shaped mushrooms are grown all year long on artificial logs. They range in size from 4 to 20 centimetres (2 to 8 inches), and have either flat or thick fleshy caps. The texture is meaty, the aroma woodsy, and the taste full-bodied and delicious.

• Enoki, sometimes called golden needle mushrooms or enokidate, are always found growing in clumps, out of a common base. These delicate mushrooms topped with tiny white caps are tall, thin and white or beige. Enoki have been described as fruity-tasting and slightly crunchy.

• Wood ear, tree ear, cloud ear or elephant ear mushrooms also grow on the bark of trees. This nearly stemless, glossy mushroom is dark brown to black and has a somewhat rubbery texture that is well known to lovers of Chinese food. A meaty mushroom, the wood ear has a damper appearance than most mushrooms but should never feel mushy.

Availability

• At one time, these extraordinary mushrooms were gathered from the wild and were available only sporadically. Now that they are cultivated, they can be found all year long. They are usually more plentiful in the spring and fall.

Selection

• Look for firm mushrooms that are in good condition and that smell good. Avoid signs of molding and dampness.

Storage

• Mushrooms are best used as fresh as possible. Oyster mushrooms may absorb strong odours from other foods in your refrigerator. Don't store these delicate mushrooms for more than a few days.

• Shiitake will keep for up to a week if they are fresh when purchased.

• Enoki must be used within a few days of purchase.

• Keep mushrooms refrigerated in a paper bag — never in plastic — or cover them with paper towels that have been slightly dampened.

Preparation

• Oyster mushrooms needn't be washed. Discard any stems that seem dense or tough. These tender mushrooms should never be served raw and are best lightly sautéed or blanched. They will shrink in size when cooked. Prepare them as you would prepare a familiar variety of mushroom, but do not overcook.

• Fresh shiitake shouldn't be washed or peeled. Simply discard the tough stems and wipe off any dirt with a damp paper towel. Fresh shiitake can be lightly cooked or eaten raw in salads. Overcooking will make the mushroom tough.

• Discard the clump at the bottom of the enoki and separate the mushrooms. There is no need to wash enoki. Be very careful not to overcook them, or they will become tough and fibrous.

• Wood ear mushrooms should be washed well. They need a longer cooking time than other types of mushrooms and are best prepared in dishes with other ingredients rather than served on their own.

Suggestions

• Grilled shiitake are excellent as an appetizer. Use shiitake in soups, salads and western or oriental dishes that call for mushrooms.

• Enoki are great raw in salads or used as a garnish; add them to a stir-fry or a dish of Chinese greens just before serving. Use uncooked enoki to garnish a clear soup. Try them in a sandwich.

• Oyster mushrooms are delicious in pasta or rice dishes. They are also superb in omelets, sauces and soups.

• Wood ear mushrooms are ideally suited to soups and stir-fries. They absorb Chinese sauces wonderfully and add a good texture to many dishes, such as casseroles, stuffings and rice dishes.

• To delight a mushroom lover, make a Chinese-style dish with shrimp and three or four kinds of oriental mushrooms.

Nutritional Value
Shiitake mushrooms
(per 100 g = 3½ oz)

			RDA**	%
calories*	(kcal)	28		
protein*	(g)	2.7		
carbohydrates*	(g)	4.4		
fat*	(g)	0.5		
fibre	(g)	0.8	30	2.6
vitamin A	(IU)	0	3300	0
vitamin C	(mg)	3	60	5
thiamin	(mg)	0.1	0.8	12.5
riboflavin	(mg)	0.46	1.0	46
niacin	(mg)	4.2	14.4	29
sodium	(mg)	15	2500	0.6
potassium	(mg)	414	5000	8.3
calcium	(mg)	6	900	0.6
phosphorus	(mg)	116	900	12.9
magnesium	(mg)	13	250	5.2
iron	(mg)	0.8	14	5.7
water	(g)	90.4		

*RDA variable according to individual needs
**RDA recommended dietary allowance (average daily amount for a normal adult)

Shiitake with tofu

(serves 4)

↓ Shiitake mushrooms

Ingredients		
12 large	shiitake mushrooms	
6	scallions (green onions)	
500 g	tofu	1 lb
45 mL	peanut oil	3 TBSP
5 mL	fresh ginger, finely grated	1 tsp
125 mL	chicken stock	1/2 cup
10 mL	dark soy sauce	2 tsp
10 mL	rice wine	2 tsp
	pinch of sugar	
15 mL	oyster sauce	1 TBSP
10 mL	cornstarch	2 tsp

Instructions

- Cut the tofu into cubes and slice the scallions into 5 centimetre (2 inch) lengths.
- In a wok, heat half the oil and sauté the tofu for about 2 minutes, turning frequently to brown all sides. Remove tofu from wok and set aside.
- Heat the rest of the oil and add the mushrooms (whole or sliced), scallions and ginger. Stir-fry for about a minute.
- Add the chicken stock, soy sauce, rice wine and sugar. Heat until bubbling, then add the tofu. Simmer a minute or two.
- Mix the cornstarch into the oyster sauce, pour it into the wok and stir until thickened.
- Serve with rice and a main course.

Oriental herbs

↑ Fresh coriander

Origin

• Herbs and spices have been used to flavour foods since mankind began to cook. Ancient texts are full of references to appetizing dishes. One Roman author deplored the vast amount of gold that was spent on spices, which were brought by caravan from the Orient. Today we are rediscovering the exotic flavours of the East.

• Chinese parsley or fresh coriander is widely used in China, Southeast Asia, India, Latin America (where it is known as cilantro) and in the Middle East. Closely related to parsley, this piquant herb originated in the Middle East and made its way to China, where it was used medicinally to treat stomach aches more than 2,000 years ago. It was probably brought by the Arabs to Europe and later taken by Spanish or Portuguese settlers to South America. It has long been very popular in Portuguese cooking and is still used in the South of France to freshen the breath after eating garlic. Its small dried seeds are essential to curries.

• Lemon grass, a herb that adds a unique flavour and fragrance to cooking, is a very important ingredient in Southeast Asia. In Indonesia, it is traditional to let young girls harvest it, as their purity is thought to improve its quality. Lemon grass, a member of the grass family, is related to most grains as well as to bamboo shoots. It contains an essential oil called citral and has long been used to make perfumes and scents. It is related to the nonedible grass that produces citronella.

• Chinese celery is not unlike ordinary celery, except for its stronger taste and pronounced fragrance. Celery is thought to have originated in Southeast Asia or the Middle East, and Chinese celery may be closer to its ancestors than western celery. The Chinese have long considered it an aid to good digestion.

• Chinese chives differ from our garden chives, and a number of different varieties are commonly used, including flowering chives. They are popular throughout the Orient.

• Other herbs and spices commonly used in oriental cuisine are chili peppers, galangal, turmeric, dill, mint, licorice basil, star anise, perilla leaves, mitsuba leaves, and all the spices associated with curry, such as cardamon, cumin, fenugreek, etc.

Varieties and description

• Fresh coriander resembles parsley. The leaf is slightly larger, flatter and much more aromatic. Some people may find it bitter or pungent at first, but it changes flavour, depending on the ingredients with which it is combined.

• Lemon grass is a long, coarse, dull green grass that becomes paler and bulbous at the root end. It exudes a sweet, lemony aroma and has a hint of ginger in its flavour.

• Chinese celery is quite small compared to western celery. The stalks are narrower and often seem spindly or a bit limp. It is almost as skimpy as the stems and leaves of celeriac. Used as a herb, it adds tremendous flavour to foods.

↑ Chinese celery

• Chinese chives or gow toy have thin, flat, dark green leaves. Another variety is the flowering chive, which has small flowers in bud or open at the very tip of the stalks. These resemble the so-called garlic chives, which are the flowering stalks of the garlic plant that taste of both garlic and chive. Another variety, called yellow chives, is grown in the dark, so its colour never develops.

Availability

• Generally available year round at Chinese or ethnic stores, except Chinese celery, which isn't always easy to find.

Selection

• Choose fresh, bright coriander, with no signs of limpness. If you are lucky enough to find it with its roots still attached, it will keep longer.

• Choose lemon grass with firm bulbs.

• Try Chinese celery to add a delicious flavour to your food, even if it doesn't look up to the standards of your normal celery.

• Look for fresh green Chinese chives that show no signs of yellowing (unless you are buying yellow chives), limpness or drying out. Flowering chives are usually fresher when the flowers are still in bud.

Storage

• Keep your bouquet of coriander in the refrigerator, either in a glass of water to prolong its life or wrapped in a paper towel inside a plastic bag.

• Cut the bulb from the stalk of the lemon grass, wrap the parts separately and store in the refrigerator for a week or two.

• Store Chinese celery and chives in the crisper for a few days.

Lemon grass ↓

Preparation

• Fresh coriander should be washed and dried just before use, then chopped finely. Use larger pieces for garnish. The root is very good in soups or stews.

• Use the thick outer stalks of lemon grass in soups or stews, then discard them after cooking. Seafood is delicious steamed or grilled on a bed of crushed stalks. Mince the succulent inner bulb and add it to soups, salad dressings, stir-fries and other dishes. You can use a mortar and pestle or the side of a broad knife to crush the bulb to release an even stronger flavour. Lemon grass goes particularly well with ginger and chili.

• Chinese celery is almost never used raw; the entire plant is chopped up and cooked.

• Chinese chives are usually minced and added to a wide variety of foods.

Suggestions

• To give a delicious coriander tang to grilled fish or chicken pieces, mix the following ingredients together: 30 mL (2 TBSP) each finely minced fresh coriander, oil, and dark soy sauce, 2 slices of fresh ginger cut in fine slivers, a squeeze of lemon and a pinch of salt and pepper. Brush half this mixture on the fish or chicken, grill the required time, then turn and brush on the rest before grilling the other side.

• Try coriander in guacamole (a Mexican avocado dip), tabouli (a Lebanese cracked wheat salad), any stuffing and many Indian dishes. It adds an excellent flavour to soups, stews, salad dressings and sauces. It is an essential herb in Mexican cuisine and complements many oriental dishes.

• Lemon grass is superb with poultry, fish and other seafood dishes. It is frequently used in marinades, Southeast Asian curries, soups and dishes that include coconut. It is even used in desserts.

• To make a tasty oriental salad, cut Chinese celery in long thin slivers, sauté in a small amount of oil and place in a bowl with a large handful of very fresh bean sprouts, 1 slivered red pepper, 4 minced scallions and a handful of snow peas cut in thin slivers. Grate in 1/2 an oriental radish and toss all ingredients with the following dressing: 30 mL (2 TBSP) light soy sauce, 10 mL (2 tsp) each sesame oil and salad oil, and a large pinch each of salt, sugar and white pepper. If desired, sprinkle the salad with sesame salt or toasted sesame seeds.

• Stir-fry slices of beef with Chinese chives and toasted peanuts. A touch of soy sauce and hoisin sauce adds a great flavor. Serve on noodles or rice. (See photo)

Chinese chives ↑

Spiced oriental eggplant

Ingredients

500 g	oriental eggplant	1 lb
6	scallions (green onions)	
30 mL	Chinese chives	2 TBSP
4	garlic cloves	
60 mL	peanut oil	4 TBSP
7 mL	fresh ginger, finely grated	1½ tsp
30 mL	dark soy sauce	2 TBSP
15 mL	hoisin sauce	1 TBSP
2 mL	Chinese chili sauce	1/4 tsp
7 mL	rice wine vinegar	1½ tsp
75 mL	chicken stock	1/3 cup
5 mL	cornstarch	1 tsp

Nutritional Value

Eggplant

(per 100 g = 3½ oz = 1/2 cup)

			RDA**	%
calories*	(kcal)	25		
protein*	(g)	1.2		
carbohydrates*	(g)	5.8		
fat*	(g)	0.2		
fibre	(g)	0.9	30	3
vitamin A	(IU)	10	3300	0.3
vitamin C	(mg)	5	60	8.3
thiamin	(mg)	0.05	0.8	6.3
riboflavin	(mg)	0.05	1.0	5
niacin	(mg)	0.6	14.4	4.2
sodium	(mg)	2	2500	0.08
potassium	(mg)	214	5000	4.3
calcium	(mg)	12	900	1.3
phosphorus	(mg)	26	900	2.9
magnesium	(mg)	16	250	6.4
iron	(mg)	0.7	14	5
water	(g)	92.4		

*RDA variable according to individual needs
**RDA recommended dietary allowance (average daily amount for a normal adult)

Instructions

- Slice the eggplant in thin rounds; removing the skin is not necessary.
- Cut the scallions in pieces of 5 centimetres (2 inches) and mince the chives.
- Heat the oil in a wok, then stir-fry the eggplant for about 2 minutes.
- Add scallions, chives, garlic and ginger; stir for another minute.
- In a small bowl, combine soy sauce, hoisin sauce, chili sauce and vinegar. Add mixture to the wok, stirring thoroughly.
- Mix the cornstarch into the stock. Add to the wok and stir until the sauce begins to thicken.
- Serve with rice and poultry, meat or fish.

Lotus

EXCELLENT SOURCE OF VITAMIN C
SOURCE OF THIAMIN AND PHOSPHORUS
AVERAGE CALORIE CONTENT

↑ Lotus root

Origin

• The lotus is a member of the water lily family. Though the plant bears beautiful flowers, its leaves don't float like Monet's water lilies but stand upright above the waterline. It originated in Southeast Asia and has been grown in ponds for at least 3,000 years. The seeds are eaten and the leaves are used for wrapping food, but it is the elongated root that is most popular outside of Asia. Its flowers, considered sacred by Buddhists, were planted for their beauty in Ancient Egypt, the Moghul pleasure gardens in Kashmir, and near many Asian temples.

Varieties and description

• The root may remind you of a string of sausages; it has a cylindrical shape that is pinched in at regular intervals. Hollow sections that run the length of each segment create a very attractive geometric lace-like design when the root is sliced. It is somewhat bland, but it has a wonderful crunchy texture and tastes good with a variety of Chinese sauces.

Availability

• Fresh lotus root can be found in the autumn.

Selection

• Choose a firm root that has no soft spots.

Storage

• Once a section is cut, it should either all be used or the unused portion should be kept in water in the refrigerator for a few days. The water must be changed daily. A whole section could be kept for a week or two.

Preparation

• Wash and peel it, then slice it thinly on the diagonal. To prevent discoloration, keep the slices of lotus root in cold water until you are ready to use them.

Suggestions

• Lotus root is an excellent addition to stir-fry.

• Garnish oriental soups with a slice of lotus root, for a decorative effect.

• Lotus root is very good in tempura.

• For a beautiful salad, blanch lotus slices, cover with boiling water and leave for about five minutes. Then drain, rinse with cold water, dry well and toss in a vinaigrette of soy sauce, Chinese vinegar, sesame oil, and a pinch of sugar and salt. They can be served as a salad alone or with other vegetables, such as red pepper, nappa or daikon.

Nutritional Value

Lotus root

(per 100 g = 3½ oz)

			RDA**	%
calories*	(kcal)	69		
protein*	(g)	2.8		
carbohydrates*	(g)	15.7		
fat*	(g)	0.1		
fibre	(g)	0	30	0
vitamin A	(IU)	0	3300	0
vitamin C	(mg)	75	60	125
thiamin	(mg)	0.14	0.8	17.5
riboflavin	(mg)	0.01	1.0	1
niacin	(mg)	0.3	14.4	2.08
sodium	(mg)	0	2500	0
potassium	(mg)	0	5000	0
calcium	(mg)	30	900	3.3
phosphorus	(mg)	103	900	11.4
magnesium	(mg)	0	250	0
iron	(mg)	0.6	14	4.3
water	(g)	—		

*RDA variable according to individual needs
**RDA recommended dietary allowance (average daily amount for a normal adult)

Oriental squash

↑ Winter melon

Origin

• Many types of squash are used in oriental cuisine. Among them are long melon, fuzzy melon, winter melon and bitter melon. Though they are called melons, they are really varieties of squash from the Cucumber family that have originated in the tropical areas of Southeast Asia.

Varieties and description

• Long melon or opo, is a long, thin, pale, yellowish-green squash, which has a thin, smooth skin. Aside from its colour, it looks like a large zucchini.

• Fuzzy melon or mogwa, is a long, thin, green squash that is covered by tiny fuzzy hairs. Some varieties are shaped like large avocados.

• Winter melon is a very large, light green squash covered in a white powdery substance, with white, mild, succulent flesh. It is usually wider than tall, and rounder than watermelon, which it resembles.

• Bitter melon, bitter gourd or balsam pear is about 12 to 30 centimetres (5 to 12 inches) long, and looks like a bumpy, grooved, light-to-dark-green cucumber. It contains a lot of quinine, which gives it a bitter or sour flavour and supposedly good medicinal properties.

Availability

• Available year long, these squash are most plentiful in the summer and fall.

Selection

• Choose both opo and fuzzy melon with firm, unblemished skin.

• Winter melon is usually sold in pieces, as a whole one can weigh up to 14 kilos (30 pounds), although some smaller varieties are sold for export. Look for firm, unbruised winter melon.

• Choose a bitter melon that is still green, as they are no good by the time they turn yellow. The dark green ones are younger and less bitter.

Bitter melon ↓

Storage

- Opo will keep in the refrigerator for at least a week.
- Fuzzy melon should be kept wrapped and refrigerated for no more than about four days.
- Wrap winter melon in plastic and refrigerate for a few days once it has been cut. If you buy a whole one, it will keep well for a few weeks stored in a cool, dry place.
- Keep bitter melon wrapped and refrigerated for up to a week.

Preparation

- Peel opo and cut according to the recipe.
- Carefully scrape off the tiny hairs from the fuzzy melon, trying not to cut into the skin, which is good to eat. Cut into long strips and stir-fry with dried or fresh shrimp and a little salt until tender. Fuzzy melon cooks quite rapidly.
- Scoop the flesh from the rind of a winter melon, remove the fibrous part which contains the seeds and use for winter melon soup or a stir-fry.

↓ Long melon or Opo

Fuzzy melon or Mogwa ↑

- Bitter melon is sometimes salted and left to stand a while, which draws out some of the bitterness. Peeling is optional. Cut it in half lengthwise, discard the seeds and the pulpy core, and slice or chop the flesh in small pieces.

Suggestions

- Opo and fuzzy melons are not dissimilar to the more familiar types of squash, and can be used in the same way.
- Winter melon is sometimes cooked in a light syrup and served as a vegetable. This is a good accompaniment to chicken dishes or spicy foods. It is frequently candied and served as a dessert.
- Bitter melon is often added to Chinese soups. It is sometimes steamed and is very good in black bean sauce. In India, it is often pickled.

Oriental squash in ginger sauce

(serves 4)

Ingredients		
1/2	bitter melon	
1/2	opo (long melon)	
2	carrots	
1	medium onion	
250 g	gai lon	1/2 lb
125 g	snow peas	1/4 lb
3-4	scallions (green onions)	
30 mL	peanut oil	2 TBSP
3	garlic cloves	
30 mL	fresh ginger, grated	2 TBSP
30 mL	oyster sauce	2 TBSP
5 mL	sugar	1 tsp
1	Thai chili pepper (optional)	
	Chinese noodles	

Instructions

- Clean and prepare the vegetables; peel, seed and chop bitter melon into small pieces, cut the opo in small chunks, the carrots into match sticks and the onion in slivers. Keep gai lon leaves whole and cut stalks into 3 or 4 pieces. Cut the scallions in about 5 pieces.
- Boil a pot of water and blanch the carrots, gai lon stems and snow peas about 2 minutes. Lift them out with a slotted spoon and drain well. In the same water, blanch the gai lon leaves for a minute, remove and drain. Then blanch the bitter melon for a minute or two, drain and discard the water.
- Cook the Chinese noodles while you prepare the stir-fry.
- In a wok, heat the oil, then add the onions, garlic and ginger. Stir for a minute, then add the scallions, opo and the other vegetables, except for the gai lon leaves. Keep stirring until the vegetables are thoroughly heated, but still crisp.
- Add the oyster sauce, sugar and chili. Mix well.
- In a large platter, place the noodles on a bed of gai lon leaves, then arrange the vegetables on top of the noodles.

Winter melon soup

(serves 4)

Ingredients		
500 g	winter melon	1 lb
125 g	cooked ham	1/4 lb
1.5 L	chicken stock	6 cups
5 mL	fresh ginger, grated	1 tsp
8-12	medium shiitake mushrooms	
15 mL	rice wine	1 TBSP
20 mL	light soy sauce	4 tsp
	salt and pepper to taste	
	slivers of bamboo shoots (optional)	

Instructions

- Peel the winter melon and discard the seeds and any stringy fibres from the seed cavity. Cut into chunks or ball shapes.
- Slice the ham in small strips.
- Heat the chicken stock and add the winter melon and the ginger. Reduce heat to a simmer and cook for about 15 minutes until tender. After the first 12 minutes, add the shiitake, whole or sliced.
- Remove soup from heat and add ham, rice wine and soy sauce. Season if necessary. Add slivers of bamboo shoots.
- Serve in a soup tureen.
- At Chinese banquets, winter melon soup is often brought to the table in a hollowed out half of a winter melon that is beautifully carved.

Chinese longbean

Varieties and description

• This amazing light or dark-coloured green bean grows from 50 to 75 centimetres (20 to 30 inches) long. They are eaten when young. Later the bean becomes fibrous and ceases to taste good, while the peas inside develop and become good to eat instead.

Availability

• Available all year, but more plentiful from late summer to fall.

Selection

• Choose light or dark green longbeans that are firm, crisp and not spongy. Generally, the thinner the bean is, the tastier it is, as once the peas inside have begun to mature, the longbean will begin to get too fibrous.

Storage

• Keep them in a plastic bag and refrigerate for no more than three or four days.

Preparation

• Wash, trim off the ends and cut on the diagonal into lengths that are easy handled, from 5 to 10 centimetres (2 to 4 inches). Longbeans adapt well to spicy sauces and are usually stir-fried rather than steamed or boiled.

Suggestions

• If you can find young, thin beans, try braiding them for a beautiful presentation.

• Blanch and drain longbeans, then stir-fry with sliced water chestnuts, garlic and a pinch of chili, sugar and salt.

• Longbeans are very good in black bean sauce or sweet and sour sauce.

• Sauté longbeans and season with hoisin sauce.

• Longbeans are good in stews.

Origin

• Also called dow gok, asparagus bean or yard long bean, Chinese longbeans from the Legume family, are more closely related to black-eyed peas than to regular string beans. Originally from Southern Asia, or possibly Africa, these beans thrive in hot climates and are grown in the Caribbean, Africa, China and other parts of Asia. To a lesser extent, they are cultivated in the U.S. for the local market.

Longbeans with Chinese BBQ pork

(serves 4)

Ingredients		
60 mL	peanut oil	4 TBSP
500 g	longbeans	1 lb
250 g	Chinese BBQ pork*	1/2 lb
15 mL	dark soy sauce	1 TBSP
15 mL	fresh ginger, finely grated	1 TBSP
15 mL	rice wine	1 TBSP
1	Thai chili pepper (optional)	

Instructions

- Prepare the beans, blanch them for 3 minutes, then drain well.
- Mix together the soy sauce, ginger and rice wine.
- In a wok, heat the oil. Add the longbeans and stir-fry about 1 minute. Add the soy sauce mixture and stir well. Remove longbeans from wok, leaving some sauce in the wok.
- Cut the Chinese BBQ pork in slices and heat quickly in the wok.
- Arrange the beans in a serving dish, with the pork on top.
- Serve with steamed rice.

Ingredients (continued)

**Available in Chinese shops*

Taro root and malanga

GOOD SOURCE OF VITAMIN A
SOURCE OF IRON
LOW FIBRE CONTENT

Origin

• These two nearly identical-looking roots are, in fact, quite different. Taro, called dasheen in the Caribbean and eddo in China, originated in Southeast Asia and India. It was taken by the Spaniards from Africa to the Caribbean, while malanga — also known as malanga amarillo, yautia, tannier and tannia — originated in the West Indies.

• These two delicious root vegetables are grown in most tropical countries. Taro has been cultivated in the Orient for more than 2,000 years and was a staple food in China long before rice was widely eaten. It is the principal ingredient in poi, the Hawaiian national dish, as well as an important food in the Caribbean and North Africa. Malanga has been a vital food in all the tropical Hispanic colonies ever since the Spaniards started foreign settlements.

Varieties and description

• There are more than 100 varieties of taro. The larger taro roots, from Thailand, the Caribbean and other countries, measure about 12 to 15 centimetres (5 to 6 inches) in length, while the smaller Chinese eddo is a quarter to half the size. These roots have brown skin with rings and white-to-cream coloured, or sometimes very pale pink, flesh. Their stalks are cooked like asparagus when fresh and their spear-shaped leaves are prepared like spinach when young and tender.

• It's very difficult to distinguish between malanga and taro; the difference lies in their taste. Malanga, popular in Latin American countries, has yellow, beige or pinkish flesh. Some varieties resemble sweet potatoes, while others look more like taro. Malanga leaves are also used like spinach.

Availability

• Taro is available year round, though the smaller variety, eddo, is mostly sold in the summer. Malanga can be found year round.

← **Taro root** (above), **Eddo** (below)

Selection

- Choose very firm, juicy, crisp roots, with no signs of soft spots or damp patches.

Storage

- Store taro as you would store potatoes, but try to use it within a week, or it may begin to go soft. Malanga should not be stored for long, whether refrigerated or at room temperature, or it will dry out and loose its distinctive nutty flavour.

Preparation

- Treat both these roots as you would treat a potato. Try them boiled, steamed, mashed or fried, and in soups, stews or casseroles. Some say these two roots are best in any dish with a rich gravy or sauce, though others say they are best fried.

- Peel and chop taro or malanga, and keep immersed in cold water until you are ready to cook it. You may want to peel the taro under running water or wear rubber gloves, if your hands are sensitive to its sticky juices, which dissipate during cooking.

- Taro should always be served hot, as its texture changes when it cools down. It can be reheated easily. When cooked, the colour may change to grayish white or mauve. If you ever had a "nest" in a Chinese restaurant, filled with seafood, it was probably made from grated, fried taro.

Suggestions

- The texture of boiled taro is like cooked chestnuts. It combines well with other ingredients to make fritters.

- Cut in large chunks and roasted with meat or poultry, basted frequently, or prepared like french fries, taro and malanga will be enjoyed by the whole family.

- One tasty way to enjoy taro is to mash about 500 g (1 lb) cooked taro with 60 mL (1/4 cup) milk, 30 mL (2 TBSP) butter or margarine, 1 large grated onion and a large handful of grated cheese. Season to taste with salt and pepper. Spread the mixture in a greased baking dish, then sprinkle on more grated cheese and paprika. Bake in a hot oven for about 20 minutes.

- Try both roots in soups or stews. Do not cook longer than necessary (about 20 minutes or so), or they may begin to disintegrate. It is best to steam or boil the roots first, then add to stews to finish cooking.

- Malanga is often boiled or puréed with butter or margarine and a little milk, salt and pepper.

Nutritional Value

Taro root

(per 100 g = 3 1/2 oz = 3/4 cup)

			RDA**	%
calories*	(kcal)	98		
protein*	(g)	1.9		
carbohydrates*	(g)	23.7		
fat*	(g)	0.2		
fibre	(g)	0.8	30	2.7
vitamin A	(IU)	820	3300	25
vitamin C	(mg)	4	60	6.7
thiamin	(mg)	0.13	0.8	16.3
riboflavin	(mg)	0.04	1.0	0.04
niacin	(mg)	1.1	14.4	7.6
sodium	(mg)	7	2500	0.28
potassium	(mg)	5.4	5000	0.1
calcium	(mg)	28	900	3.1
phosphorus	(mg)	61	900	6.7
magnesium	(mg)	0	250	0
iron	(mg)	1	14	7.1
water	(g)	73		

*RDA variable according to individual needs
**RDA recommended dietary allowance (average daily amount for a normal adult)

Bird's nest made from taro ↓

Chinese okra

↑ Chinese okra

• Also called luffa, angled luffa, silk melon or sing-kwa, this odd vegetable, which is really a type of squash from the Cucumber family, grows on a tropical vine and is only edible when picked very young. It is thought to come from India originally. About 30 centimetres (1 foot) long, this heavily ridged, dull green vegetable is very attractive and extremely unusual.

• The tips of its ten sharp edges are peeled off, then the luffa is cut into stick shapes or thickish slices of at least a centimetre (about 1/2 an inch), so that they maintain their shape when cooked. Do not overcook. It is very tasty when prepared in a tempura batter, stir-fried, prepared in oyster sauce or in soups. If it is very fresh, you can slice it thinly into a salad. Luffa has become a popular vegetable in the Caribbean.

Chinese okra with shrimp

(serves 4)

Ingredients		
1 or 2	Chinese okra	
1	bunch of gai lon	
1	bunch of scallions (green onions)	
45 mL	peanut oil	3 TBSP
5 mL	fresh ginger, finely grated	1 tsp
3	garlic cloves, minced	
500 g	medium shrimp	1 lb
15 mL	rice wine	1 TBSP
30 mL	lemon grass, finely minced	2 TBSP
5 mL	sesame oil	1 tsp
45 mL	oyster sauce	3 TBSP

Instructions

• Cut the okra in rounds or sticks, the gai lon into lengths of 5 centimetres (2 inches) and mince the scallions.

• In a wok, heat the peanut oil. Add the ginger, garlic, scallions and shrimp. Stir for about 2 minutes.

• Add the rice wine, lemon grass, okra and gai lon and stir constantly for a few minutes, until the gai lon is tender.

• Gradually add the oyster sauce and sesame oil as you stir. If necessary, add a small amount of water.

• Serve with rice or Chinese noodles.

Green papaya and green mango

Green papaya ↑

↓ Green papaya salad

• When underripe, both fruits are used extensively in Southeast Asian and Indian cuisine. The green fruit is treated more as a vegetable and is often included in spicy dishes.

Green papaya salad

(serves 4)

Ingredients		
1	large green papaya	
1/2	nappa	
125 mL	unsalted cashews, toasted	1/2 cup
	juice of 1 lemon	
30 mL	Chinese fish sauce	2 TBSP
3-4	garlic cloves, minced	
2.5 mL	salt	1/2 tsp
15 mL	sesame oil	1 TBSP
1	starfruit, sliced	
	watercress to garnish	

Instructions

• Put aside a few whole nappa leaves to line your platter or bowl and shred the rest.

• Peel and halve the papaya, and discard the seeds. Grate it by hand or in a food processor.

• Mix the lemon, fish sauce, garlic and oil and use most of it to mix into the shredded nappa.

• Arrange the shredded nappa and the papaya on whole nappa leaves. Sprinkle on the cashews and garnish with starfruit slices and watercress. Drizzle on the remainder of the dressing.

Green mango curry

(serves 4)

Ingredients		
3	green mangos	
190 mL	grated fresh coconut	3/4 cup
5 mL	crushed coriander seeds	1 tsp
10 mL	fresh ginger, finely grated	2 tsp
1	Thai chili pepper, minced	
500 mL	coconut milk	2 cups
2	small onions	
30 mL	oil	2 TBSP
	a pinch of turmeric	
2 to 2.5 mL	curry powder	1/4-1/2 tsp
2	bay leaves	
15 mL	fresh coriander	1 TBSP
16 to 20	medium shrimp, cooked	
	handful of bread crumbs	

Green mango ↑

↓ Green mango curry

Instructions

- Peel the mangos, remove pits and cube the flesh. Grate the coconut, reserving the milk.
- Mix the grated coconut with the crushed coriander seeds, ginger, chili and enough coconut milk to make a paste.
- Mince the onion and sauté it over low heat until transparent. Add the curry, turmeric, bay leaves and coconut paste and continue to cook for 4 to 6 minutes.
- Add the mangos, the fresh coriander and the rest of the coconut milk; let simmer for about 10 minutes. If you are using fresh shrimp, add them when you add the mangos.
- Add the cooked shrimp and the bread crumbs, then remove from the heat. Garnish with a sprig of fresh coriander and serve with rice.

↑ Curuba

Nutritional Value

Passion fruit

(per 1 medium = 18 g = 2/3 oz)

			RDA**	%
calories*	(kcal)	18		
protein*	(g)	0.4		
carbohydrates*	(g)	4.2		
fat*	(g)	0.1		
fibre	(g)	2	30	6.7
vitamin A	(IU)	126	3300	3.8
vitamin C	(mg)	5	60	8.3
thiamin	(mg)		0.8	
riboflavin	(mg)	0.02	1.0	2
niacin	(mg)	0.3	14.4	2.08
sodium	(mg)	5	2500	0.2
potassium	(mg)	63	5000	1.26
calcium	(mg)	2	900	0.2
phosphorus	(mg)	12	900	1.3
magnesium	(mg)	5	250	2
iron	(mg)	0.29	14	2
water	(g)	13.1		

*RDA variable according to individual needs
**RDA recommended dietary allowance (average daily amount for a normal adult)

Availability

- Passion fruit is available year round, as there are cultivators all over the world. The season in Thailand is from August to December; in New Zealand, from February to June; in Brazil, from January to July. They are most plentiful from March to September.

Selection

- Smooth-skinned passion fruit will need ripening. Wrinkled fruit is ripe and juicy on the inside.

Storage

- Keep passion fruit at room temperature until the skin becomes wrinkled and shrivelled. When the fruit looks as though it should be thrown out, it will taste best. It can then be refrigerated until used. A little mold on the skin won't hurt, providing the skin is not cracked, but the fruit should be consumed immediately at that point. Both the flesh and the juice can be frozen for later use.

- Maracuya, a variety of large yellow-orange passion fruit with white and gray flesh, is sweeter than most. It is usually eaten when the skin is still smooth.

Preparation

- The usual way to eat a passion fruit is to cut it in half and scoop out the flesh with a spoon. Some people like to sprinkle on a little sugar.

Suggestions

- Passion fruit juice is exceptionally good; briefly blend the pulp and seeds alone, with water or with fruit juice, strain through a very fine strainer or muslin cloth to discard the seeds and chill. Sweeten if desired. Serve plain or in a tropical cocktail, such as a daiquiri.

- Add 1 or 2 passion fruit to fruit salad or punch.

- The pulp of just 1 or 2 fruit is often enough to flavour a sauce or gravy, or even a cake filling, pastry cream, crêpe topping, etc.

- The tart flavour lends itself particularly well to sorbet and other frozen desserts.

Cheese cake with passion fruit topping

Ingredients		
	Crust	
190 mL	flour	3/4 cup
60 mL	ground almonds or hazelnuts	1/4 cup
60 mL	sugar	1/4 cup
5 mL	grated lemon peel	1 tsp
	pinch of salt	
5 mL	vanilla	1 tsp
1	egg yolk	
60 mL	soft butter	1/4 cup
	Cake ingredients	
500 g	smooth fresh cottage cheese	1 lb
250 g	cream cheese	1/2 lb
250 g	sour cream	1/2 lb
190 mL	sugar	3/4 cup
45 mL	flour	3 TBSP
10 mL	grated lemon peel	2 tsp
10 mL	grated orange peel	2 tsp
5 mL	vanilla	1 tsp
5-6	eggs	
60 mL	heavy cream	1/4 cup
	Passion fruit topping	
4-6	passion fruit	
10-15 mL	sugar	2-3 tsp
2.5 mL	unflavoured gelatin	1/2 tsp
60 mL	water	1/4 cup

Instructions

- To make the crust, mix the first 5 ingredients together, then add the rest of the crust ingredients. Mix thoroughly, then refrigerate for 1 hour.
- Preheat oven to 200 °C (400 °F) and grease a 20 to 22 centimetres (8 or 9 inches) springform pan. Coat the bottom of the pan with part of the crust mixture (you will have to use your fingers to spread it) and bake 8-10 minutes or until it is golden. Coat the sides of the pan with the rest of the mixture, then refrigerate until you are ready to bake the cake.
- Preheat the oven to 260 °C (500 °F) while you prepare the next step.
- In a large bowl or food processor, combine all the cake ingredients except the eggs and the cream. Beat until well blended.
- Beat in the eggs, one at a time, then add the cream and beat until smooth.
- Pour the mixture into the springform pan and bake for 10 minutes.
- Reduce the heat to 120 °C (250 °F) and bake for another hour.
- Cool on a rack, then very carefully remove the sides of the springform pan.
- Refrigerate for 3 hours.
- Glaze with passion fruit topping: Mix the pulp of 4 to 6 passion fruit with the sugar, then add the gelatin dissolved in hot water. When it has almost set, spread it evenly across the top of the cake, then refrigerate until it has become firm.

Granadilla ↓

Prickly pear

LOW IN CALORIES
VERY GOOD SOURCE OF VITAMIN C
EXCELLENT SOURCE OF MAGNESIUM

Origin

• This refreshing, juicy, desert fruit is also known as an Indian pear or fig, Barbary fig, cactus pear, tuna and sabra. Indigenous to Mexico and the Southwestern United States, the cactus pear was extremely important to pre-Columbian civilization, as a source of water for people and their animals. The fruit grows well on its own, with no fertilizers or pesticides, in arid or semi-arid regions.

• In moderation, the seeds of this fruit have excellent digestive properties, though some people find them too hard and prefer to discard them. The Indians and the early Spanish settlers drank juice made from the cactus fruit to bring down a fever. Spread around the world by the Spaniards, prickly pears are now grown in many South American countries, including Brazil, Chile, Argentina and Peru, Mediterranean countries, especially Italy, Spain and Israel, and in many parts of the United States. Central America, Australia, large areas of Africa and Asia also have prickly pears. Mexico still remains the world's largest producer, consumer and exporter.

Varieties and description

• Egg-shaped or pear-shaped, the average fruit measures 8 to 10 centimetres (3 to 4 inches) in length and is covered with sharp spines that are removed before the fruit is sold. The skin of different varieties of prickly pears ranges from green to yellow, orange, pink or crimson, while the flesh, which contains numerous small hard edible seeds, can be green, yellow or red. The flavour of this thirst-quenching, juicy fruit is comparable to sweet, mild watermelon.

• The large flat "leaves" or pads of one variety of cactus, called nopales, are a popular vegetable in Mexico. These leaves are light green and tender but crisp. They are usually harvested in the spring, before the cactus flowers.

• Pitahaya, which is the fruit of another type of cactus found in Colombian tropical woodlands, is sometimes available in specialty shops in the winter and late spring. This beautiful little fruit turns red or yellow when ripe and has pale flesh dotted with tiny dark seeds. It is known as an excellent natural laxative.

Availability

• Cactus fruit is available year round from Argentina and Brazil, and seasonally from other countries — June to October from Mexico, August or September to December from the United States (Arizona, California, Washington and Oregon), in the late autumn from the Mediterranean countries and February to April from Chile.

Selection

• Choose firm, unblemished fruit. Moldy spots may indicate soft, squishy, unappetizing flesh.

Storage

• If necessary, let fruit ripen at room temperature until it is slightly soft and uniform in colour. (Some fruit can be juicy inside when still hard outside.) It can then be refrigerated for a few days. This fruit is very perishable when fully ripe. Prickly pears must be well ripened to be appetizing. If the flesh of a fruit seems dry, it is best to discard it.

• Cactus leaves, nopales or nopalitos, have a very short shelf-life; use as soon as possible after purchase.

Preparation

- The fruit tastes best slightly chilled. Most prickly pears can be handled with no problem, although some tiny, nearly invisible barbed hairs may remain behind after the fruit has been de-spined. Carefully cut a thin slice off each end, then make a skin-deep incision from one end to the other. Use fingers or a fork and knife to peel back the skin completely and remove the fruit.
- To prepare nopales, cut out the thorny "eyes" and peel off the edge with a sharp knife. Slice into long stick shapes, then boil or steam about 10 minutes until tender, treating them like green beans. In Mexico, they are usually boiled with a pinch of baking soda, salt and a small minced onion. Nopales are often served with scrambled eggs, omelets or stews.

Suggestions

- Prickly pears are most often eaten thinly sliced, with a splash of lemon or lime to enhance the flavour, on their own or in fruit, vegetable, chicken or shrimp salads.
- As with most tropical fruits, they make delicious jam, preserves and sorbet. It is best to strain out the seeds for sorbet.
- An attractive fruit plate can be created with slices of the differently coloured cactus fruit.
- For a healthy breakfast, with cereal if desired, mix small chunks of prickly pear into plain yogurt, sweetened with a spoonful of honey.

Nutritional Value

Prickly pear

(per 1 medium = 103 g = 3 1/2 oz)

			RDA**	%
calories*	(kcal)	42		
protein*	(g)	0.8		
carbohydrates*	(g)	9.9		
fat*	(g)	0.5		
fibre	(g)	1.9	30	6.3
vitamin A	(IU)	53	3300	1.6
vitamin C	(mg)	14	60	23.3
thiamin	(mg)	0.01	0.8	1.3
riboflavin	(mg)	0.06	1.0	6
niacin	(mg)	0.5	14.4	3.5
sodium	(mg)	6	2500	0.2
potassium	(mg)	226	5000	4.5
calcium	(mg)	58	900	6.4
phosphorus	(mg)	25	900	2.7
magnesium	(mg)	88	250	35
iron	(mg)	0.31	14	2.2
water	(g)	90.2		

*RDA variable according to individual needs

**RDA recommended dietary allowance (average daily amount for a normal adult)

Prickly pear cocktail

(serves 2)

Ingredients		
3-4	prickly pears, carefully peeled	
125 mL	orange or grapefruit juice	1/2 cup
	squeeze of lemon	
5 mL	honey	1 tsp
	vodka or tequila to taste	

Instructions

- The colour of this drink will depend on the variety of prickly pear.
- Blend the first 4 ingredients until smooth, then strain out the seeds.
- Chill if desired, add vodka and serve with ice.
- This drink is also refreshing with no alcohol. Try blending in a ripe banana and a scoop of ice cream instead.

Persimmon

1. Hachiya
2. Fuyu

EXCELLENT SOURCE OF VITAMIN A
VERY GOOD SOURCE OF VITAMIN C
CONTAINS POTASSIUM AND MAGNESIUM

Origin

• The persimmon, originally from China, though cultivated and improved by the Japanese for more than 1,000 years, is the only well-known fruit from the large ebony family of plants, very few of which produce any edible fruit. It is also called kaki, which is part of its scientific name. The Portuguese brought back persimmon seeds from Japan in the 16th century and later introduced them to Brazil and several West Indian islands. Commodore Perry brought persimmon seeds to America from Japan in 1855, and others have introduced different varieties since then.

• The persimmon tree is considered a beautiful ornamental and its wood is good for carving and fine furniture. The Japanese recommend persimmons for fatigue and hangovers, probably because they are so nutritious.

• Major producers today — aside from China and Japan, where it is considered to be the national fruit — include California, Brazil, Chile, Spain, France, Italy and Israel. New Zealand and Australia are now producing in greater quantities.

Varieties and description

• Though hundreds of varieties of this beautiful fruit exist, only two principal types are well known — the hachiya and the fuyu. Both come in various sizes, similar to apples or tomatoes.

• Hachiyas are bright heart-shaped and orange-red inside and out. One of the sweetest fruits in world, this popular variety (and other soft varieties) should only be eaten when fully ripened and therefore very soft, as the tannic acid dissipates when the fruit is ripe. Some tannin may remain in the skin, which is usually not eaten unless the fruit is completely soft. Hachiyas have a few small black seeds in the centre, which can be eaten.

Physalis

EXCELLENT SOURCE OF VITAMINS A AND C
GOOD SOURCE OF FIBRE AND IRON
VERY LOW IN CALORIES

Origin

- The most sought after species of physalis are the Cape gooseberry, the ground cherry and the tomatillo. Common names for the tomatillo include Mexican husk tomato and jamberberry or jamberry, while other species of physalis are sometimes called strawberry tomato, golden berry, poha and tipari. The Chinese lantern plant or winter cherry is a well known type of physalis grown principally as an ornamental; its fruit is not particularly good, but its bright orange lantern-like calyxes are very attractive.

- Physalis are members of the Solanaceæ family, along with tomatoes, peppers and many other vegetables. Botanists are not certain whether they originated in South America (probably Peru) or in Japan and China. They grow wild all over the world, including the Americas, India, Africa, Hawaii and the Orient, in all sorts of climates and soil conditions. Ground cherries were well known to the native peoples of North and South America before the arrival of Europeans.

- Physalis peruviana became naturalized more than 200 years ago in South Africa, hence its common name, Cape gooseberry. From there it was introduced to Europe.

- New Zealand has become a major producer-exporter of Cape gooseberries. Ground cherries are grown on a small scale in Quebec, Ontario, throughout New England and in many other areas of the United States.

- Tomatillos, probably native to Mexico, are mainly available there, though a limited supply can be found in California and other states where fresh Mexican food is popular.

Varieties and description

- The name physalis comes from the Greek word "physa", meaning bladder, which refers to the enlarged, straw-coloured, papery, lantern-like calyx that encloses the fruit. Cape gooseberries and ground cherries are beautiful little yellow berry-like fruits that have juicy flesh, full of very small, tender, edible seeds.

- Both Cape gooseberries and ground cherries can be very sweet with a slightly tart aftertaste when fully ripe, or quite sour when underripe. Cape gooseberries average 2 centimetres (3/4 inch) in diametre, while ground cherries are usually smaller and tomatillos are larger.

- Tomatillos are sharper tasting than other varieties of physalis. Whereas most physalis is treated like fruit, tomatillos are used as vegetables. They look like large green cherry tomatoes. If left to over-ripen, they will turn yellow or purplish.

Availability

- The season in New Zealand for Cape gooseberries is from March to June. The season in Africa and Asia is from December to June, though they are not usually imported from there. North American ground cherries are available only for a short period in the summer.

- Tomatillos grow year round, but are rarely seen in Canada or the Northern United States.

Selection

- Look for firm, unblemished, ripe fruit, with their calyxes in good condition.

Storage

- Don't peel the husks off or wash the fruit until you are ready to use them. Keep them at room temperature if they are to be used within a day or two. They keep well when refrigerated for a week or more.

- Tomatillos will keep well for several weeks in the refrigerator. Once they are cooked, they can be frozen.

Preparation

- Physalis, with their husks pulled back, look beautiful used as an edible garnish. To cook them, simply remove husks and wash them. If you find that they become too liquidy when cooked, use a very little cornstarch, tapioca or flour to thicken.

- Tomatillos have a sharper flavour when uncooked, and are tasty sliced into sandwiches or salads. They should be husked just before use, well washed and the small hard core is usually removed. They can be served as a side vegetable, simply grilled with a little salt and pepper. Tomatillos are a basic ingredient in Mexican cuisine, especially in "salsa verde", the famous Mexican green sauce that accompanies almost everything. This sauce can be used uncooked; when cooked, it has a milder taste and a longer shelf life.

➠ Suggestions

• Add ripe uncooked Cape gooseberries or ground cherries to a fruit salad, or serve them as an hors d'oeuvre with a tangy dip.

• Physalis is attractive and delicious used to top open-faced fruit tarts or to fill fancy pastry.

• Use them alone or combined with apples or pears in pie or crumble.

• Simmered with a little sugar and a touch of ginger, if desired, physalis makes a wonderful tangy sauce to serve with roast duck or chicken, as well as with pork or lamb dishes. This sauce is equally good over ice cream, fruit salad and with other desserts.

• Cape gooseberries and ground cherries make excellent jam or preserves.

• To make a basic tomatillo salsa, simply chop up or purée tomatillos with fresh cilantro, minced onion, a clove of garlic, lime or lemon juice, a pinch of sugar, salt and pepper, and chili pepper if desired. Delicious with avocados and cucumbers, certain soups, grilled chicken, broiled fish, egg and cheese dishes and many other foods. Serve cold, whether fresh or cooked.

Nutritional Value

Physalis — ground cherries

(per 140 g = 5 oz = 1 cup)

			RDA**	%
calories*	(kcal)	74		
protein*	(g)	2.7		
carbohydrates*	(g)	15.7		
fat*	(g)	1.0		
fibre	(g)	3.9	30	13
vitamin A	(IU)	1008	3300	30.5
vitamin C	(mg)	15	60	25
thiamin	(mg)	0.2	0.8	25
riboflavin	(mg)	0.06	1.0	6
niacin	(mg)	3.9	14.4	27
sodium	(mg)		2500	
potassium	(mg)		5000	
calcium	(mg)	13	900	1.4
phosphorus	(mg)	56	900	6.2
magnesium	(mg)		250	
iron	(mg)	1.4	14	10
water	(g)	119.6		

*RDA variable according to individual needs
**RDA: recommended dietary allowance (average daily amount for a normal adult)

Physalis shortcake

(serves 4)

Ingredients		
250 mL	physalis	1 cup
30-45 mL	sugar	2-3 TBSP
60 mL	Grand Marnier or Cointreau	1/4 cup
190 mL	whipping cream	3/4 cup
	pinch of sugar	
4	individual-sized sponge cakes	

Instructions

• Cut each physalis in half. Place halves in a bowl with the sugar and liqueur and allow to macerate for a few hours.

• Just before serving, whip the cream with the pinch of sugar and spread evenly over the little sponge cakes.

• Drain physalis, reserving the liqueur to add to your coffee. Spoon the fruit over the top of the cakes and serve.

Quince

GOOD SOURCE OF VITAMIN C
LOW IN CALORIES, VERY LOW IN SODIUM
GOOD RANGE OF MINERALS

Origin

- Quince is a member of the Rosaceæ family and is related to apples, pears, plums, cherries and many other fruits. It is native to the western part of Asia, from Turkey to Iran. Well known as "golden apples" to the Greek and Roman civilizations, quince has long been a symbol of love and happiness. It was part of the wedding ritual, as the fruit was dedicated to Aphrodite and Venus. It may even have been the biblical fruit of the tree of knowledge.

- Its Portuguese name, marmelo, is the origin of the word marmalade, which was made with quince long before it was made with oranges. In France, especially in the area near Orleans, it has been used for hundreds of years to make a sweet called cotignac and to flavour brandy. Quince was once far more popular and commercially important than it is today, because of its very high pectin content. Since artificial pectin was invented, quince has greatly diminished in importance. It was also used to treat digestive disorders. At one time, because of its gelling properties, it was used by hairdressers in preparations concocted to hold the shape of elaborate hair-dos!

- Quince was introduced to England in the 16th century and to North America in the 17th century. Quince trees were once very common in gardens throughout New England. Today, quince is grown in many parts of the world, including the Mediterranean area, Latin America (where it is called membrio), the Middle East and the United States.

Varieties and description

- Round or pear-shaped, downy or smooth-skinned, all varieties have a lovely, fruity fragrance. One variety, pineapple quince, is about three times the size of a normal quince.

- The fruit turns from green to yellow as it ripens and bruises very easily, often leaving blotchy areas on the surface that don't really affect the quality of the fruit. The yellowish flesh, which turns dark when exposed to the air, is sour, very firm and dry. The core is similar to an apple core, but it has a lot more pips. Quince fruits are inedible raw, yet delicious when cooked.

Availability

- The main season for quince from Europe and North America is from September or October to December or January. South American quince is available in April and May.

Selection

- Choose large fruit, with some yellow in the skin to show that they have been picked ripe. Bruised fruit are sometimes unavoidable, and are just as good to use.

Storage

- Quince will keep well for several weeks, providing they are not too badly bruised. Wrap them individually for protection and refrigerate.

Preparation

- Wash well and peel if desired. Removing the core and seeds requires a sharp knife and should be done before cooking. If you are preparing a lot of quince, keep the cut pieces in cold water with lemon juice to prevent them from discolouring. Sweeten to taste after cooking, unless you are making jam that requires sugar in its initial preparation. The flesh turns dull pink or red when cooked. Ginger compliments the flavour of quince.

Suggestions

- Quince sauce, made the same way as apple sauce, is excellent with roast lamb and pork or poultry dishes. Use all quince, or half apples half quince.

- Quince baked or stewed with raisins, honey and vanilla and served with cream is a simple yet outstanding dessert.

- The most popular use of quince is for jams and jellies.

- Leave a quince out or hang it to spread a fragrant scent in a room or closet.

- Quince do not fall apart when cooked, the way apples or pears do, and are therefore very good in such dishes as stews that require a long cooking time.

- A quince added to an apple pie or crumble improves the flavour and aroma immeasurably.

- Quince "cheese" is a well known treat in Europe and is delicious served with apples or pears; make quince jam and continue cooking it until it is extremely thick. Transfer it to a loaf pan, and dry it out in a warm oven or an airing cupboard until it sets solid. Cut off thin slices as required.

- Quince syrup is great on pancakes, ice cream and fruit salad.

Moroccan lamb casserole with quince and dates

(serves 4)

Ingredients		
500 g	**lean lamb, cubed**	**1 lb**
1	**dozen dry shallots**	
2-3	**carrots, cut in matchsticks**	
1	**chicken stock cube**	
2	**garlic cloves, finely minced**	
2-2.5 mL	**salt**	**1/4-1/2 tsp**
	pinch of pepper	
	pinch of paprika	
	pinch of cayenne	
45 mL	**fresh coriander, minced**	**3 TBSP**
5-10 mL	**fresh ginger, finely grated**	**1-2 tsp**
5 mL	**fresh mint, finely chopped**	**1 tsp**
2	**quince**	
1	**dozen pitted dates**	
60 mL	**raisins**	**1/4 cup**
15 mL	**flour**	**1 TBSP**

Instructions

- In a casserole with a cover or an electric slow-cooker, put all but the last 4 ingredients, with enough water to barely cover the meat.

- Put the casserole into a 230 °C (450 °F) oven. When the liquid begins to bubble, turn the heat down to 160 °C (325 °F) and continue to cook for another 1 1/2 hours.

- Peel, core and slice the quince. Cut the dates in half lengthwise.

- Remove a small amount of liquid from the casserole and mix with the flour until a smooth paste is formed. Add this paste, along with the quince, dates and raisins to the casserole. Return to the oven for another 30 minutes. When it is done, add salt and pepper to taste.

- Serve with couscous or rice, and fresh peas or beans.

- Mint sauce is a good condiment for this dish.

Quince compote

(serves 4)

Ingredients		
4	quince	
5 mL	fresh ginger, finely grated	1 tsp
60-125 mL	sugar	1/4-1/2 cup
1	cinnamon stick	

Instructions

- Wash, peel and core the quince and remove all the seeds. Then cube it.

- In a medium saucepan, add about 1.25 centimetres (1/2 an inch) of water, the quince and the ginger. Bring to a boil, then simmer until tender, about 15 to 20 minutes.

- After it has cooked, add sugar to taste and a cinnamon stick. Refrigerate when cool.

Nutritional Value

Quince

(per 1 medium = 92 g = 3 1/4 oz)

			RDA**	%
calories*	(kcal)	53		
protein*	(g)	0.4		
carbohydrates*	(g)	14.1		
fat*	(g)	0.1		
fibre	(g)	1.6	30	5.3
vitamin A	(IU)	38	3300	1.2
vitamin C	(mg)	14	60	23
thiamin	(mg)	0.02	0.8	2.5
riboflavin	(mg)	0.03	1.0	3
niacin	(mg)	0.2	14.4	1.4
sodium	(mg)	4	2500	0.2
potassium	(mg)	181	5000	3.6
calcium	(mg)	10	900	1.1
phosphorus	(mg)	16	900	1.8
magnesium	(mg)	7	250	2.8
iron	(mg)	0.64	14	4.6
water	(g)	77.1		

*RDA variable according to individual needs
**RDA recommended dietary allowance (average daily amount for a normal adult)

Loquat

EXCELLENT SOURCE OF VITAMIN A
LOW IN CALORIES AND SODIUM
CONTAINS POTASSIUM AND MAGNESIUM

Origin

- A member of the Rosaceæ family, the loquat is related to the apple, pear, cherry and plum. It is sometimes known as the Japanese plum or the Japanese medlar, though it isn't a true medlar and it didn't originate in Japan. It is native to southern China and its name derives from the Cantonese word "lu-kwyit", which means rush-orange. The Japanese have cultivated loquats for many centuries and have greatly improved them.

- The attractive loquat tree, prized by violin makers for its beautiful wood, was introduced to France in the early 19th century as an ornamental, but the fruit was not eaten. It later became a popular fruit throughout the Mediterranean area because it ripened earlier than other spring fruits. In the Orient, this fruit was once used to treat coughs and colds.

- Today it grows in many sub-tropical areas throughout the East, in India, Taiwan, most Mediterranean countries especially Italy, Spain and Israel, in Central and South America, the West Indies, as well as in California and Florida. Chile and Brazil are among the biggest producers in South America.

Varieties and description

- These small, pear-shaped fruits, which grow in clusters, measure from 4 to 8 centimetres (1 1/2 to 3 inches) in length. The fruit's thin, smooth, shiny skin, which ranges from pale yellow to deep orange, is sometimes covered with a whitish down. The flesh, juicy and tender but crisp, ranges in colour from cream to deep orange. Each fruit contains several glossy, inedible pips.

- The sweet taste is reminiscent of cherries and plums. The flavour differs in sweetness according to the variety but can be very acidic if underripe. There are about 10 varieties of loquats, some of which have a lovely, fruity aroma.

Availability

- Brazilian and Chilean loquats are available from August to October and oriental loquats are in season in April and May. Loquats from Florida are available from March to May, while those from California are available from March to June.

Selection

- This fruit bruises extremely easily, but it will still taste good. In Europe, people are more likely to buy them when covered in brown spots, as this indicates perfect ripeness and therefore better flavour.

Storage

- Loquats are picked ripe and ready to eat. Use as soon as possible, as they are highly perishable. If they are not too spotted, they can be kept for a day or two at room temperature; otherwise, they should be refrigerated.

Preparation

- The delicious loquat is usually eaten fresh, alone or in a fruit salad. It should be washed, cut in half, then the seeds must be removed. The peel comes off easily if preferred, though it is edible. If cooking loquats, add a few seeds to the preparation, then discard them; this enhances the flavour and helps the fruit set or the sauce thicken.

Suggestions

- Poached in a light syrup and complimented by a hint of cinnamon or ginger, loquats are delicious served at any time of the day. Try them with cream.

- They are a perfect fruit for pies, open-faced tarts, or used to fill fancy pastries.

- Loquats make excellent jellies or preserves. Try to use less-ripe fruit for this purpose.

Nutritional Value

Loquat

(per 100 g = 3 1/2 oz)

			RDA**	%
calories*	(kcal)	47		
protein*	(g)	0.4		
carbohydrates*	(g)	12.1		
fat*	(g)	0.2		
fibre	(g)	0.5	30	1.7
vitamin A	(IU)	1528	3300	46.3
vitamin C	(mg)	1	60	1.6
thiamin	(mg)	0.02	0.8	2.5
riboflavin	(mg)	0.02	1.0	2
niacin	(mg)	0.2	14.4	1.4
sodium	(mg)	1	2500	0.04
potassium	(mg)	266	5000	5.3
calcium	(mg)	16	900	1.8
phosphorus	(mg)	27	900	3
magnesium	(mg)	13	250	5.2
iron	(mg)	0.28	14	2
water	(g)	86.7		

*RDA variable according to individual needs
**RDA recommended dietary allowance (average daily amount for a normal adult)

Tamarillo

LOW IN CALORIES AND SODIUM
EXCELLENT SOURCE OF VITAMINS A AND C
GOOD SOURCE OF MINERALS

Origin

• These attractive subtropical fruits were called tree tomatoes until 1967, when New Zealand growers renamed them tamarillos as part of their campaign to popularize them. They are a member of the large Solanaceae family, which includes many popular vegetables, notably tomatoes, potatoes, peppers and eggplant. There are not too many well-known fruits in this family, though the various species of physalis have become more sought after in the last few years.

• Originating in the Peruvian Andes, tamarillos are now grown all over Central and South America, notably Colombia, Venezuela, Ecuador and Brazil, the Caribbean, some regions of Africa, India, Australia and Southeast Asia. New Zealand, one of the major producers, is working on an improved, sweeter tamarillo.

Varieties and description

• There are two principal varieties of tamarillos. The most common one is a deep, rich red-to-purple and the slightly sweeter variety is a beautiful golden yellow-orange. Tamarillos are egg-shaped, about 5 to 10 centimetres (2 to 4 inches) long, with both ends tapering to a point. They are usually sold with a thin piece of hard stem still attached. The bitter, inedible, thin skin is glossy and smooth. The deeply coloured flesh contains small flat seeds that are very similar to tomato seeds. Tamarillos have an unusual, tart, tangy taste and are used as a vegetable as well as a fruit.

Availability

• New Zealand tamarillos are available from April to October, while those from South America can be found year round.

Selection

• Choose unblemished tamarillos that are still firm, but yield to gentle pressure. Unripe tamarillos can be very bitter.

Storage

• If they are hard, leave at room temperature for a few days. If ripe, wrap and keep them in the crisper for up to a week or more.

Preparation

• Tamarillos taste best when sweet and sour (i.e. sugar and lemon) ingredients are added to them. In South America, many people just cut them in half, sprinkle them with salt or sugar and scoop them out with a spoon. Tamarillos are an acquired taste, and most North Americans will prefer them cooked.

• Cooked or raw, they need to be peeled. Blanch them for a minute or two in boiling water and the skins will peel off easily.

Suggestions

• The easiest way to enjoy a tamarillo is to peel and slice it, spread the slices out flat, sprinkle with sugar and lime or lemon juice and a little rum if desired; let macerate for a few hours or bake them about 15 to 20 minutes. Serve with ice cream. This is also tasty served with grilled meat or chicken.

• Tamarillo is excellent in chutney and sweet and sour dishes.

• South Americans use it to make jam, juice and ice cream.

• To use tamarillos as a vegetable, sprinkle slices with lemon, salt and pepper and grill or sauté them for a few minutes. Serve with meat, fish or poultry.

• A puréed tamarillo will add tang and beautiful colour swirled into cake frosting, soufflé, apple pie and many other desserts.

• Tamarillo slices can be marinated for an hour or two in a spicy vinaigrette and used to decorate a salad platter.

Nutritional Value

Red tamarillo

(per 100 g = 3 1/2 oz)

			RDA**	%
calories*	(kcal)	36		
protein*	(g)	2		
carbohydrates*	(g)	5.3		
fat*	(g)	0.6		
fibre	(g)	3.9	30	13
vitamin A	(IU)	1865	3300	56
vitamin C	(mg)	31	60	52
thiamin	(mg)	0.05	0.8	6.3
riboflavin	(mg)	0.01	1.0	1
niacin	(mg)	0.2	14.4	1.4
sodium	(mg)	1.6	2500	0.06
potassium	(mg)	310	5000	6.2
calcium	(mg)	11	900	1.2
phosphorus	(mg)	39	900	4.3
magnesium	(mg)	21	250	8.4
iron	(mg)	0.6	14	4.3
water	(g)	87		

*RDA variable according to individual needs
**RDA recommended dietary allowance (average daily amount for a normal adult)

Tamarillo parfaits

(serves 4)

Ingredients		
2	tamarillos	
	juice of 1/2 orange	
	juice of 1 lemon	
1	ripe banana	
15 mL	liquid honey	1 TBSP
15-30 mL	kirsch or other liqueur	1-2 TBSP
190 mL	whipping cream	3/4 cup
	pinch of sugar	
	vanilla ice cream	

Instructions

• Peel tamarillos after blanching them in hot water for a minute.

• Set aside 4 very thin slices of tamarillo for garnish. Place them in a plate and sprinkle with lemon juice and sugar. Refrigerate for an hour or two.

• In a blender, combine the tamarillos, orange and lemon juice. Blend until smooth. Strain out seeds if the tamarillo is too tart.

• Add banana, honey and the liqueur; blend until smooth.

• Refrigerate for at least 2 hours.

• When ready to serve, whip the cream with a pinch of sugar. In 4 parfait glasses, alternate layers of vanilla ice cream, tamarillo sauce and whipped cream. Top each serving with whipped cream sliced tamarillo and a fancy biscuit.

Pomegranate

GOOD SOURCE OF POTASSIUM
SOURCE OF VITAMINS C AND B
VERY LOW IN SODIUM

Origin

• The pomegranate, a unique fruit in a family of its own (Punicaceæ), probably originated from Persia to Afghanistan. It has been cultivated from Southern Europe to China and Japan, as well as in Northern Africa for thousands of years. Its name, derived from Latin, means apple of numerous seeds. The Persians used it to dye fabric and leather and it has long been used medicinally by herbalists to treat a multitude of inflammations, from sore throats to rheumatism. It was even used in former times to massage and rejuvenate aging skin.

• The pomegranate is a symbol of fertility in many cultures, notably in ancient Greece. According to the prophet Mohammed, eating pomegranates will cleanse the soul of envy. The Arabs supposedly introduced pomegranates to North Africa, although they have been found in ancient Egyptian tombs, indicating their presence long ago. The Moors brought pomegranates to Spain, from where they spread throughout Europe. Frequently mentioned in the Old Testament, they were often used by the ancient Hebrews as an ornamental motif in their architecture.

• Today pomegranates grow in the sub-tropics all over the world, including Thailand, South America, the Middle East, the entire Mediterranean area, the Caribbean and California, where Spanish missionaries introduced them about 200 years ago.

Varieties and description

• There are only slight differences between the few varieties of pomegranates — usually variations in the colour of the skin and the sweetness of the fruit. The size and shape of a large orange or a small grapefruit, the pomegranate is covered with a thin, tough, smooth yellow-to-bright red or purplish skin that protects the fruit inside, keeps it juicy, and gives it a long shelf life. Opposite the stem end is a prominent, stiff calyx. Inside, dozens of small seeds enveloped by juicy bright cranberry-red flesh are clustered in six compartments made of bitter, white, inedible membrane. The unusual flavour of the flesh is sweet, yet tart and very refreshing.

Availability

• Pomegranates are available from September to December, with peak supplies in October.

Selection

• For a juicy pomegranate, choose an unblemished fruit that feels heavy for its size. The larger the fruit, the better developed and sweeter it is.

Storage

• Keep pomegranates at room temperature for a few days, or refrigerate for a week or two — possibly longer if the fruit is in good condition. The seeds can be removed from the fruit and frozen. Pomegranate juice can also be frozen.

Preparation

• Simply cut the fruit in quarters to get at the seeds, but be careful not to stain yourself with the juice. A tidier but much slower method is to lightly score the skin, peel it off, then peel off the white pith to reveal the seeds.

Whichever way you eat it, it takes a long time to remove all the seeds. Your patience will be rewarded.

- Some people just enjoy the fruit and spit out the seeds, while others eat the small seeds as well. Pomegranates can be eaten plain or sprinkled with sugar or salt to taste.

- Pomegranates have been used for centuries to make grenadine syrup.

Suggestions

- To make pomegranate juice — a delicious drink on its own or mixed with other juices — remove all the seeds, then use a juice extractor or a blender and a sieve. Try freezing some pomegranate juice in an ice cube tray, for later use in orange juice or exotic drinks.

- Sprinkle a handful of pomegranate seeds over fruit, cottage cheese or vegetable salads to create a beautiful ruby-jeweled effect. Desserts, such as ice cream, trifle and crêpes, as well as main courses, from omelets to grilled fish, also look attractive garnished this way.

- Pomegranate sorbet is refreshing and appealing.

- Pomegranate syrup is excellent as a sauce for other fruits, such as baked apples or poached pears, or used in drinks. Mash the fruit lightly with a fork and let macerate with sugar (half sugar to fruit) about 1 day. Boil, strain out pips, bottle and refrigerate.

- Pomegranate juice is a wonderful addition to a sweet and sour sauce.

Nutritional Value

Pomegranate

(per 1 medium = 154 g = 5 1/2 oz)

			RDA**	%
calories*	(kcal)	104		
protein*	(g)	1.5		
carbohydrates*	(g)	26.4		
fat*	(g)	0.5		
fibre	(g)	0.3	30	1.0
vitamin A	(IU)		3300	
vitamin C	(mg)	9	60	15
thiamin	(mg)	0.05	0.8	6.3
riboflavin	(mg)	0.05	1.0	5
niacin	(mg)	0.5	14.4	3.5
sodium	(mg)	5	2500	0.2
potassium	(mg)	399	5000	7.9
calcium	(mg)	5	900	0.6
phosphorus	(mg)	12	900	1.3
magnesium	(mg)		250	
iron	(mg)	0.46	14	3.3
water	(g)	124.7		

*RDA variable according to individual needs

**RDA recommended dietary allowance (average daily amount for a normal adult)

Arabian chicken in pomegranate sauce

(serves 4)

Ingredients

30 mL	oil	2 TBSP
4	chicken legs or breasts	
1	medium onion	
	salt and pepper to taste	
1	chicken stock cube	
	juice of 2-3 pomegranates	
8	dried pears	
	juice of 1/2 lemon	
15 mL	honey	1 TBSP
15 mL	flour	1 TBSP

Instructions

- In a large heavy skillet, heat the oil and brown the chicken all over.

- Chop the onion and add it to the chicken. Sauté until the onion is browned. Sprinkle on a little salt and pepper.

- Add about 75 mL (1/3 cup) of water to the pan and the stock cube. Boil, then cover and simmer.

- Add the pomegranate juice, the dried pears (sliced in halves or quarters lengthwise), the lemon juice and the honey.

- Continue cooking on low heat for about an hour, turning the chicken over 2 or 3 times.

- Combine the flour with a little of the stock and use it to thicken the sauce. Decorate each plate with a small handful of pomegranate seeds, if desired.

- Serve with rice or cracked wheat (bulgar) and spinach or Swiss chard.

Feijoa

(pronounced fee-jo-a)

VERY LOW IN CALORIES
GOOD SOURCE OF VITAMIN C AND FIBRE
VERY LOW IN SODIUM

Origin

• Also called pineapple-guava though it is not a true guava, this subtropical fruit is one of the few edible fruits in the Myrtle family, which also includes the guava. The feijoa was named after J. da Silva Feijo, who was the director of the Natural History Museum in San Sebastian, Brazil, more than a century ago.

• The feijoa originated in a region of South America that includes southern Brazil, Uruguay, Paraguay and northern Argentina. It was introduced by a French horticulturalist to Europe toward the end of the last century, where it was grown as an ornamental for its beautiful blood-red flowers. Brought to New Zealand at about the same time for the same reason, it is only in recent years that it has been cultivated for its fruit.

• New Zealand is one of the largest producers of feijoa today. The fruit is also now beginning to be grown commercially in California, where it was first introduced at the turn of the century. Other producers include Israel and the Soviet Union.

Varieties and description

• This green, egg-shaped, smooth-skinned fruit is usually 5 to 8 centimetres (2 to 3 inches) long and has a prominent calyx. The skin, thin yet tough and very bitter, is considered inedible. The sweet, tangy, aromatic, cream-coloured flesh has the granular consistency of a pear. When sliced crosswise, the feijoa offers an attractive pattern formed by its slightly gelatinous centre, which contains some tiny, soft, edible seeds.

• The feijoa exudes a sublime tropical, fruity aroma. Its taste can vary enormously, not only from country to country and from tree to tree, but from one fruit to the next. Agricultural scientists are constantly working to produce improved and more uniform fruit.

Availability

• New Zealand feijoas are available from March to June, while those from California can be found from September to November.

Selection

• Choose smooth-skinned, unblemished, aromatic fruit.

Storage

• Feijoas are picked ripe and shipped by air. They may soften slightly if left at room temperature; if that happens, they should be eaten as soon as possible. Refrigeration is not normally necessary.

Preparation

• Peel and slice, or cut in half and scoop out the flesh with a spoon. Sliced feijoas may turn brown after being exposed to the air, so it is best to squeeze on a little lemon or lime juice. They are more attractive sliced crosswise or on the diagonal. If used in cooking, remove the peel first.

Suggestions

• Feijoas make excellent jam, as well as a delicious sorbet or mousse.

• Sprinkle sliced feijoas with a little brown sugar and lemon or orange juice mixed with a little vanilla. A hint of ginger will add an agreeable flavour. Chill well. Use a sprinkling of Grand Marnier, Calvados or Port if desired. Serve with ice cream, yogurt or sour cream.

• Feijoas can be substituted for or combined with apples or bananas in your favourite recipes. Try a few added to an apple pie for a wonderful flavour and aroma.

• For an exotic drink, float a slice of feijoa in a glass of champagne.

• Try marinating and cooking whole or sliced feijoas in your favourite barbecued pork recipe.

• Feijoa fritters make a delightful dessert.

Feijoa crumble

(serves 6)

Ingredients		
Filling		
	zest and juice of 1 orange	
45 mL	brown sugar	3 TBSP
2.5 mL	cinnamon	1/2 tsp
2 mL	dry ginger	1/4 tsp
10	feijoas, peeled and sliced	
60 mL	raisins	1/4 cup
Crumble topping		
190 mL	granola cereal	3/4 cup
190 mL	unbleached flour	3/4 cup
75-125 mL	slivered almonds	1/3-1/2 cup
2.5 mL	cinnamon	1/2 tsp
30-45 mL	brown sugar	2-3 TBSP
	pinch of salt	
125 mL	softened butter or margarine	1/2 cup

Instructions

• To make the filling: In a bowl, sprinkle the orange juice over the brown sugar and half the orange zest. Mix in the cinnamon and ginger and stir thoroughly.

• Peel and slice the feijoas in rounds, exposing the pretty pattern. Add to the orange and sugar mixture, and toss gently until well mixed in. Add the raisins and transfer to a baking dish, arranging the fruit attractively.

• To make the crumble topping: Mix all the dry ingredients with the rest of the orange zest in a bowl. Cut the butter or margarine into this mixture with two knives, until the butter is in pea-sized bits. Then, using your fingers, rub the butter thoroughly into the flour and granola. (This crumble topping is good for a variety of fruits, including rhubarb, mango, cherries, peaches, apples and pears. Feijoas combine well with apricots, apples and pears.)

• Cover the fruit evenly with the crumble topping, and bake in a preheated oven at 200 °C (400 °F) for about 20 to 30 minutes. The top should be golden brown.

• Serve with ice cream.

Nutritional Value

Feijoa

(per 100 g = 3½ oz)

			RDA**	%
calories*	(kcal)	34		
protein*	(g)	0.8		
carbohydrates*	(g)	9.5		
fat*	(g)	0.3		
fibre	(g)	4.3	30	14
vitamin A	(IU)	100	3300	3
vitamin C	(mg)	27	60	45
thiamin	(mg)	0	0.8	0
riboflavin	(mg)	0.01	1.0	1
niacin	(mg)	0.2	14.4	1.4
sodium	(mg)	3	2500	0.1
potassium	(mg)	130	5000	2.6
calcium	(mg)	4.8	900	0.5
phosphorus	(mg)	12	900	1.3
magnesium	(mg)	7	250	2.8
iron	(mg)	0.1	14	0.7
water	(g)	85		

*RDA variable according to individual needs
**RDA recommended dietary allowance (average daily amount for a normal adult)

Horned melon

VERY LOW IN CALORIES AND SODIUM
VERY GOOD SOURCE OF VITAMIN C
CONTAINS IRON AND POTASSIUM

Origin

• The outlandish horned melon is definitely a conversation piece! It is one species of more than 1,000 members of the Cucurbitaceæ family, which also includes melons and squash. Originally from Southwest Africa, it is also known as the African horned cucumber or jelly melon. This fruit was brought to New Zealand about 60 years ago, and until recently was grown mainly for its decorative appeal. In the last 10 years, New Zealand growers have developed it commercially for export.

Varieties and description

• Horned melons are usually about 10 to 12 centimetres (4 to 5 inches) long, and weigh 250 to 400 grams (8 to 14 ounces). The fruit is harvested when its skin begins to show yellow or orange streaks and is fully ripe when it changes to bright orange. The brighter the hue, the better the flavour. Its hard skin is covered with pronounced spikes and its attractive, juicy, emerald green flesh is filled with soft edible seeds, reminiscent of cucumber seeds. The taste is often described as a combination of cucumber, lime, banana and melon.

Availability

• Horned melon is harvested from February to April and is usually available from late February to June.

Selection

• As long as a horned melon shows yellow or orange colour in its skin, it will ripen properly. It should feel firm and have unblemished skin.

Storage

• Horned melons have an extremely long shelf life — up to several months if handled properly! They should be kept at room temperature, and should not be refrigerated unless they are to be prepared within a day of refrigeration. They are best consumed within two weeks of purchase. Do not keep horned melon with fruit that gives off a lot of ethylene gas, such as bananas or apples, or its shelf life will be shortened.

Preparation

• Slice the fruit in half or quarters lengthwise and scoop out the pulp. The skin is inedible, but half shells make very decorative serving receptacles.

• Some people prefer the flesh with a little sugar sprinkled over it.

Suggestions

• Use horned melon in a centrepiece or fruit basket.

• To make an exotic drink, blend the pulp and strain it. Add a squeeze of lemon or lime juice, and sugar to taste. (Melt the sugar in a little hot water.) Add a little orange or melon liqueur if desired. Serve in tall glasses filled with ice.

• Use the pulp of horned melon to create a beautiful salad dressing for seafood, poultry or vegetable salads.

• Horned melon flesh and melon balls, macerated in a mixture of liqueur, citrus zest and a little sugar, served in the half shell, looks and tastes wonderful.

Nutritional Value

Horned melon

(per 100 g = 3 1/2 oz)

			RDA**	%
calories*	(kcal)	24		
protein*	(g)	0.9		
carbohydrates*	(g)			
fat*	(g)			
fibre	(g)	0.6	30	2
vitamin A	(IU)	250	3300	7.6
vitamin C	(mg)	23	60	38
thiamin	(mg)		0.8	
riboflavin	(mg)		1.0	
niacin	(mg)		14.4	
sodium	(mg)	1	2500	0.04
potassium	(mg)	250	5000	5
calcium	(mg)	20	900	2.2
phosphorus	(mg)	21	900	2.3
magnesium	(mg)		250	
iron	(mg)	1	14	7.1
water	(g)			

*RDA variable according to individual needs
**RDA recommended dietary allowance (average daily amount for a normal adult)

Horned melon with shrimp salad

(serves 4)

Ingredients		
2	horned melons	
250-500 g	medium-sized shrimp, cooked	1/2-1 lb
1	stalk celery, finely chopped	
1	small cucumber, finely chopped	
45-60 mL	mayonnaise	3-4 TBSP
15 mL	fresh dill, finely chopped	1 TBSP
15 mL	fresh chives, finely chopped (or 2 scallions, finely chopped)	1 TBSP
	juice of half a lemon	
	salt and pepper to taste	

Instructions

- Cut the horned melons in half and scoop out the pulp.
- Combine the shrimp salad ingredients and add between 1/3 and 1/2 of the horned melon pulp.
- Fill the horned melon shells with the shrimp salad, and serve.
- Use the rest of the horned melon pulp for an exotic drink or add it to a fruit salad.

Guava

EXCELLENT SOURCE OF VITAMIN C
VERY LOW IN CALORIES AND SODIUM
HIGH IN VITAMIN A AND FIBRE

Origin

• Guavas are presumed to have originated in an area stretching from Mexico and the West Indies to Brazil. Spanish explorers brought the fruit back to Europe and later introduced it to their colonies in the East. A member of the myrtle family, guavas are related to cinnamon, cloves, eucalyptus and many other fragrant woods and spices. Only a very few edible fruits, such as the feijoa, the water rose and the rose apple, are produced by members of this family, of which the guava is the best known. Traditionally, ripe guava has been used as a laxative. However, unripe ones can have the opposite effect.

• Today they can be found in tropical and sub-tropical regions around the world. Brazil and Taiwan are among the biggest suppliers.

Varieties and description

• The shape and size of large or comice pears, the most popular types of guava average 8 centimetres (3 inches) in diametre, though some varieties are smaller or larger. The edible skin, ranging from thin to thick, is usually pale green or yellow. The flesh colour can be white, yellow, pink or red, depending on the variety. The central part contains from few to many small hard seeds that are edible, though too gritty for some. Guavas vary from wonderfully fragrant to musky; the aroma from one fruit will pervade a whole household. They taste sweet and exotically tropical.

• There are more than 150 varieties of guavas, only a few of which are seen in North America. One variety rarely seen in North America is the delicious small Brazilian egg-sized yellow or red strawberry guava, which tastes vaguely like strawberries. One type of Oriental guava is hard, green-skinned, white-fleshed and as crunchy as an apple.

Availability

• Brazilian and Thai guavas can be found year round, while Chilean guavas are available only from April to August. Guavas from Taiwan are imported from August to April. Florida and Mexico have two harvests — from January to March and from June to October.

Selection

• Choose unblemished fruit, either slightly soft for immediate use or firm for later use.

Storage

• Ripen guavas for a few days at room temperature, until they give slightly to gentle pressure. Refrigerate only when ripe. Although the appearance is not spoiled with long storage, the flesh can become mealy.

Preparation

• Peel if desired and halve or slice. When preparing guava nectar or jam, cook the entire fruit and strain out the seeds afterwards. Guava jelly is excellent, but usually does not set solidly. It has a texture more like liquid honey.

• The fruit can be puréed when uncooked then strained and incorporated into a great variety of recipes, from such desserts as sorbet, ice cream, mousse and pudding, to sauces for chicken or pork, or drinks like guava milk shakes.

Suggestions

• A sliced guava will add a sunny, tropical taste to fruit salad.

• Try poaching guavas in honey and rum or Southern Comfort with a squeeze of lemon or lime. Serve with yogurt or cream.

• Slices of grilled guava are outstanding with meat dishes.

• Scoop out the centre of ripe guavas and stuff them with cottage cheese, ice cream, fruit salad or custard.

• Guava paste, made by boiling guavas and sugar down to a thick consistency, is great spread on bread with cream cheese, or eaten with stronger cheeses such as Roquefort or sharp cheddar.

Nutritional Value

Guava

(per 1 medium = 90 g = 3 1/4 oz)

			RDA**	%
calories*	(kcal)	45		
protein*	(g)	0.7		
carbohydrates*	(g)	10.7		
fat*	(g)	0.5		
fibre	(g)	5	30	16.7
vitamin A	(IU)	713	3300	21.6
vitamin C	(mg)	165	60	275
thiamin	(mg)	0.05	0.8	6.3
riboflavin	(mg)	0.05	1.0	5
niacin	(mg)	1.1	14.4	7.6
sodium	(mg)	2	2500	0.08
potassium	(mg)	256	5000	5.1
calcium	(mg)	18	900	2
phosphorus	(mg)	23	900	2.5
magnesium	(mg)	9	250	3.6
iron	(mg)	0.28	14	2
water	(g)	77.5		

*RDA variable according to individual needs
**RDA recommended dietary allowance (average daily amount for a normal adult)

Muffins with guava jelly

(makes 12-14 muffins)

Ingredients		
125 mL	bran *(not bran cereal)*	1/2 cup
125 mL	toasted wheat germ	1/2 cup
250 mL	whole wheat flour	1 cup
12 mL	baking powder	2 1/2 tsp
2.5 mL	baking soda	1/2 tsp
2.5 mL	salt	1/2 tsp
125 mL	chopped walnuts	1/2 cup
1	egg	
190 mL	guava jelly*	3/4 cup
190 mL	milk	3/4 cup
60 mL	melted margarine	1/4 cup
12-14	walnut or pecan halves for garnish	

Instructions

• Preheat the oven to 200 °C (400 °F). Use medium or large paper baking cups inside a muffin tin, or grease the muffin tin well.

• In a large bowl, mix the first 7 ingredients.

• In another bowl, combine the egg, guava jelly, milk and melted margarine. Pour the liquid ingredients into the dry ones, gently stirring just enough to moisten all the mixture. (Don't over-mix or the muffins won't rise properly.)

• Fill the paper baking cups 3/4 full, garnish the top of each one with half a walnut, and bake for about 20 minutes. Pierce a muffin with a toothpick; if it comes out clean, with no batter sticking to it, the muffin is ready. Cool before serving.

* guava jelly

• Use half ripe and half unripe guavas to get the right pectin balance. Slice the guavas, then cook them in a little water until they are soft. Use just enough water to prevent them from sticking to the pot, and add small amounts, as needed, during cooking.

• Place the cooked guava in a strainer set over a bowl; use a spoon to mash the fruit, keeping all the liquid and pulp that passes through the strainer and discarding the seeds.

• Measure the cooked fruit and add an equal amount of sugar and the juice of half a lemon. Boil this mixture until the consistency appears right for jelly; i.e., when you dip a spoon into it, the drops should run into each other instead of dripping separately. When cooled, it should have the consistency of honey.

• Serve leftover jelly with the muffins or ice cream. It keeps well for several weeks in the refrigerator.

Cherimoya and soursop

↑ Soursop

CHERIMOYA IS A FAIRLY GOOD SOURCE OF FIBRE AND VITAMIN B
SOURSOP IS AN EXCELLENT SOURCE OF VITAMIN C

Origin

• Cherimoyas originated in the highlands of Peru and Ecuador and were once the prized fruit of the Incas. Mark Twain described this wonderful fruit as "deliciousness itself" and it truly is one of the most luscious, delectable tropical fruits in the world.

• A member of the Annona family, it is closely related to the soursop or guanabana, the sugar apple or sweetsop and the netted custard apple or bullock's heart. Many annona fruits are commonly referred to as "custard apples", adding to the list of confusing names. European settlers helped spread these fruits around the globe in the 18th century but, until recently, they have remained virtually unknown in most of North America.

• The cherimoya, a delicate fruit that requires time-consuming hand pollination, was introduced to California about 100 years ago, but it is only in recent years that production has begun to expand. It is also grown in Brazil, Argentina, Chile, Peru, Ecuador, Mexico, Spain, Israel, Thailand, Indonesia, Australia and New Zealand. The atemoya, now grown in Florida, Australia, Israel and South Africa, is a cross between a cherimoya and a sweetsop.

• Soursops, probably originally from Colombia, are now grown throughout South and Central America and in most Caribbean Islands, as well as some African and Southeast Asian countries. Soursops have been used medicinally for centuries, to treat stomach and intestinal complaints.

Varieties and description

• The heart-shaped cherimoya, about the size of a small grapefruit, has an inedible skin covered by smooth, flat scales similar to those of an immature pine cone. Its skin ranges from bronze to jade green. Inside, its creamy, succulent, white flesh, which has the texture of custard, is dotted with inedible shiny black seeds. Its unique, mouth-watering taste and aroma is like a mixture of every rich tropical fruit plus some secret ingredient. (Native South Americans used to crush the seeds and soak them in water to use as an insect repellent!)

• Atemoyas are similar in appearance and taste, but have more pronounced, droopy scales.

• The larger soursop, weighing from 500 grams to 5 kilos (1 to 11 pounds), has a lumpy, kidney shape. Its bright green skin is covered with protruding scales or spines that are slightly prickly. Its sweet and sour flesh is creamy white and somewhat granular, at times mealy. It contains numerous inedible black seeds.

Availability

• Peruvian cherimoyas are in season from June to October, Californian ones from December to May, and Chilean ones in October and November. The Floridian atemoya can be found in August and September.

• Soursops are available year round, though they are not often seen in North America, except in communities with large populations of West Indians, South Americans or Orientals.

Selection

• These delicate fruits are picked when they are still firm and shipped by air. Though their skin looks tough and leathery, it is thin and bruises easily; they are often packed individually in protective foam netting. They can also be damaged by cold temperatures. Small surface scars do not indicate spoiled fruit, but large discoloured patches of skin may indicate overripened fruit.

Storage

• Cherimoyas and other annona fruits should be left at room temperature until they yield to gentle pressure, then they should be refrigerated and eaten within a few days. Don't let them get mushy. Never refrigerate any of these fruits before they have fully ripened but don't allow them to overripen or they will begin to ferment. The skin, especially of soursops, may darken slightly as they ripen.

Preparation

• Cherimoyas and atemoyas lose a lot of flavour if cooked; these distinctive fruits are most enjoyable when chilled and scooped straight out of the shell. Halve or quarter each fruit and remove the central fibre only if it seems tough. Discard the seeds.

• Soursops are usually consumed as juice or ice cream, though a few varieties are sweet enough to eat as is.

Suggestions

• Cherimoyas are almost always eaten plain, though they can be used for exotic drinks, fruit salad, mousse, parfait, ice cream or sorbet.

• Squeeze a little fresh orange juice over the cherimoya, to enhance its slight vanilla flavour, or serve with orange slices.

• Annona fruits are exceptional in trifle.

• Unless the soursop is sweet enough to eat as is, it is best to discard the skin and seeds, then purée or press through a strainer for use in drinks, sorbet and other frozen desserts. As you strain it, keep pouring water and lime juice or milk over the fruit pulp to help you extract all of its juices. Add a little sugar or maple syrup to taste and a hint of nutmeg or cinnamon if desired.

• Soursop is also good mixed with mashed banana and cream and sweetened to taste. Try a soursop and banana milkshake.

Nutritional Value
Cherimoya
(per 100 g = 3 1/2 oz)

			RDA**	%
calories*	(kcal)	94		
protein*	(g)	1.3		
carbohydrates*	(g)	24		
fat*	(g)	0.4		
fibre	(g)	2.1	30	7
vitamin A	(IU)	10	3300	0.3
vitamin C	(mg)	8.9	60	14.9
thiamin	(mg)	0.1	0.8	12.5
riboflavin	(mg)	0.1	1.0	10
niacin	(mg)	1.3	14.4	9
sodium	(mg)		2500	
potassium	(mg)		5000	
calcium	(mg)	23	900	2.5
phosphorus	(mg)	40	900	4.4
magnesium	(mg)		250	
iron	(mg)	0.5	14	3.6
water	(g)	74		

*RDA variable according to individual needs
**RDA recommended dietary allowance (average daily amount for a normal adult)

↓ Cherimoya

Cherimoya daiquiri

(serves 2)

Ingredients		
1	cherimoya	
75 mL	light rum	1/3 cup
30 mL	curaçao	2 TBSP
15 mL	lime juice	1 TBSP
5-10 mL	sugar	1-2 tsp
500 mL	crushed ice	2 cups
	starfruit or orange slices to garnish	

Instructions

- Cut the cherimoya in half; scoop out the flesh and discard all the seeds.
- Put the pulp in a blender with the other ingredients, and blend until the liquid is thick, frosty and smooth.
- Garnish the glasses with starfruit or orange slices.

Cherimoya ice

Ingredients	
1	cherimoya
2	oranges
	sugar to taste

Instructions

- Cut the oranges in half and scoop out the flesh.
- Put the orange and cherimoya pulp in a blender. Add sugar.
- Blend to desired consistancy. Fill the orange halves with the blended fruits.
- Freeze until firm 4 to 6 hours.
- Garnish with **fresh mint leaves.**

Cardoon

Origin

• Cardoon — also known as cardoni or cardone — originated in the Mediterranean area. It is the ancestor of the globe artichoke and both are edible thistle members of the large Compositæ family, which also includes lettuce, dandelions, endives, salsify and Jerusalem artichokes. The word cardoon derives from the Latin word for thistle.

• Well liked by the Romans and their descendants, cardoon became widespread throughout Southern Europe by the Middle Ages. It was consumed in large quantities by pregnant women, who believed that eating cardoon would give them male children.

• Cardoon is mainly grown in Italy, France, Spain, Argentina and Australia. Early colonists and Italian settlers in North America first cultivated this unusual winter vegetable on the East Coast and later in California.

Varieties and description

• There are several varieties of cardoon, which differ slightly in colour. Shaped somewhat like celery, with long rather flexible stalks, it is blanched to a grayish green colour a few weeks before harvesting by being covered with heaped earth, cardboard or dark plastic.

• Its outer stalks are fibrous and covered with soft prickles. Its inner stalks are lighter coloured and more tender and succulent.

• The large top leaves are always cut off when harvested, which discolours the cut edge. The flavoursome taste is like a combination of artichoke, celery and salsify.

Availability

• Cardoon is available from September to March, mostly in Italian and specialty shops.

Selection

• Look for the most firm and crisp cardoon.

Storage

• Wrap the base in a damp paper towel, place in a paper or plastic bag and refrigerate in the crisper for no more than a week or two, depending on its condition. Don't let it dry out.

Preparation

• Cardoon cannot be eaten raw. It is often cooked before being added to a recipe to tenderize it and remove most of its bitterness. If using only half a cardoon, cut it in half crosswise and reserve the bottom half for another meal. Slice the browned ends off the top, discard the tough outer leaves (unless the grower has done so), then separate and wash the rest. Cardoon may discolour as you work with it, but the colour will even out once it is cooked.

• Trim the leafy edges off each stalk. Use a vegetable peeler to take off the thin layer containing celery-like strings from the back of all but the inner stalks. Cut the stalks in stick shapes, or according to your recipe. Add the cardoon to a large quantity of boiling salted water with a little vinegar or lemon juice and cook for about 30 minutes, or until it is tender. Drain and discard the cooking water.

Suggestions

• Cardoon is delicious pre-cooked, then baked au gratin in mornay sauce. It is also good braised, creamed or served simply with a little melted butter, seasoned with salt and pepper.

• Sauté garlic and onions in a little olive oil, add sliced plum tomatoes, cooked cardoon, fresh thyme or oregano, salt and black pepper. Simmer for 5 to 10 minutes and serve as a vegetable side dish.

• Make a zesty vinaigrette, toss well with cooked cardoon slices and refrigerate overnight. Add it to a Mediterranean-style salad, garnished with olives or capers.

• Try cardoon in Mediterranean dishes, especially soups or stews.

Cardoon tempura with anchovy sauce

(serves 4-8)

Ingredients		
1/2-1	bunch cardoon	
250 mL	ice water	1 cup
1	egg	
250 mL	flour	1 cup
	pinch of salt	
	vegetable oil for frying	
	Anchovy sauce	
60 mL	sour cream	4 TBSP
1	tin anchovies, including the oil	
30 mL	salad oil	2 TBSP
45 mL	white wine vinegar	3 TBSP
5 mL	fresh thyme, minced	1 tsp
15-30 mL	fresh parsley, minced	1-2 TBSP

Instructions

- Cut cardoon into stick shapes, about 5 centimetres long by 2 centimetres wide (2 inches by 3/4 inch).
- Add to boiling salted water with a little vinegar or lemon juice and cook for about 30 minutes, or until it is tender. Drain, then dry thoroughly.
- Heat the vegetable oil to about 175 °C (350 °F). When the oil is nearly hot and you are ready to begin cooking, make the batter.
- Mix the ice water and egg in a bowl. Add the flour and salt, barely mixing the ingredients together so that the batter remains light.
- Dip the vegetable sticks in the batter a few at a time, then cook in the hot oil for a few minutes, until lightly browned.
- Remove the cooked pieces with a slotted spoon and drain off excess oil on paper towels.
- Blend the sauce ingredients thoroughly, using the oil from the anchovies as well. Sprinkle the parsley over the top.
- Serve with anchovy dipping sauce.

Greens

(Kale / Collards / Dandelion)

↑ Dandelion

EXCELLENT SOURCE OF VITAMINS A AND C
VERY LOW IN CALORIES AND SODIUM
HIGH IN FIBRE, CALCIUM AND MAGNESIUM

Origin

• A number of green vegetables, including kale, collards, dandelion, turnip greens, beet greens, Swiss chard, spinach and mustard greens are collectively called greens. Many greens have come to be thought of as food for the poor, but the truth is that they are extremely nutritious and should form part of a healthy diet.

• Kale and collards, which are closely related, are among the oldest members of the Crucifer family, along with cabbage, kohlrabi, broccoli and cauliflower. Dandelions are members of the Compositæ family, along with sunflowers, lettuce, chicory, endives, salsify and artichokes. Other greens are in the Chenopod (goosefoot) family.

• Kale (also known as borecole) and collards originated from Eastern Europe to Western Asia and probably have been known to man for thousands of years. It is difficult to be sure of their history, as numerous types of cabbage were described by the Greeks and Romans, but the exact translation of their names remains questionable.

• Kale has long been widespread as a healthy, hearty winter vegetable in Europe. The Scots are perhaps its biggest fans, followed by Germans and Scandinavians. It was among the first of the European plants brought to North America by colonists. Collards spread from Europe to Africa centuries ago and were brought to North America by slaves. It has remained a popular vegetable ever since, in the American South.

• Dandelions are thought to have originated in Western Asia or the Middle East. For many centuries they have been used medicinally as a diuretic and as a remedy for kidney problems. During the Renaissance, they were considered beneficial for the complexion. Early Dutch settlers probably were the first to use wild dandelions in North America; they enjoyed them in salad with a sweet hot cider vinegar dressing. The French probably eat the most dandelions — they blanch them, with methods similar to those used in cultivating endives, to obtain a white, sweeter and more tender leaf. North Americans have only recently begun to cultivate dandelion leaves for the specialty market.

Varieties and description

• Kale looks like parsley on a giant scale. Among its many varieties, differences occur in the degree of curliness of the leaves, in colour, which ranges from light to dark green and even bluish green, in the length of the stems and leaves and especially in the taste (some are more cabbage-like than others). There also are quite a few attractive ornamental varieties. Kale tastes sweetest when harvested after being exposed to frost. Its flavour is like that of mild cabbage.

• Collard leaves are large, smooth and dark green, with ruffled edges. The tough central rib is usually discarded. It is milder tasting than kale, its closest relative.

• The word dandelion comes from the French "dent-de-lion", meaning lion's tooth, so-named for the jagged leaves that edge the long, slender stalks. The leaves and stalks are either fresh green or white, in the case of blanched leaves. They have a very palatable, pleasantly bitter flavour. Cultivated varieties show vast improvements in taste and size over wild dandelions.

Availability

• Kale grows year round, but is most abundant from December to April. October to May is the main season for collards, though it can sometimes be found at other times of the year. September to March is the season for white or blanched dandelion and March to December for green varieties.

Selection

• Young greens are always more mild and tasty. Choose fresh-looking leaves, avoiding limp or yellowed ones. When leaves are still attached to the roots, they stay fresher for a longer time. Remember that fresh greens usually cook down to about 1/4 of their volume, so always get lots.

Storage

• Keep unwashed greens refrigerated in the crisper. Wrap them in a damp paper towel and place in a plastic bag. They are best used within a few days of purchase while still fresh and at their peak.

Preparation

• Most greens compliment each other and can be mixed together or substituted for one another in recipes. Kale and collards require very thorough washing, especially kale, whose crinkly leaves can hide sand or soil. Soak in tepid water, then run them under the tap. Strip off and discard the central stalks from each kale or collard leaf, unless they are very young and tender. Kale and collards should be sliced in thin ribbons for use in soup or more coarsely sliced for sautéeing.

• Blanch the leaves in boiling water and dry them completely before sautéeing. Alternatively, pan fry first, then add a little stock, cover and simmer until tender. They can be left whole or torn into large pieces for boiling, steaming or microwaving. Collards are most appetizing when cooked in a seasoned broth for about 10 to 12 minutes, while kale will cook well in half that time. Add the broth to a soup.

• Cultivated dandelion leaves need to be washed and well dried. They are most often used in salads or cooked like spinach.

Suggestions

• In the American South, collards and kale are often prepared with bits of crispy bacon or diced, smoked ham. They are frequently added to soups or stews, especially those containing beans and barley or spicy sausages.

Collards ↓

• Kale and collards can be treated like spinach. Cook them lightly and use as a bed for shepherd's pie or poached eggs or as a filling for omelettes or crêpes.

• Greens can be cooked simply and served with a little butter, a squeeze of lemon and salt.

• For a tasty gourmet salad, coarsely chop dandelion leaves, radicchio, arugula, watercress and bibb lettuce. Add some thinly sliced sweet onion rings and dress with olive oil, red wine vinegar, a garlic clove, a touch of Dijon mustard, a pinch of sugar, salt and freshly ground black pepper. Add anchovies, croutons or slivered sweet peppers for an extra touch.

• Dandelion leaves are delicious boiled or steamed for a few minutes until tender in water or a little soup stock, drained, then served with butter and salt, possibly with a squeeze of lemon or oil and vinegar. Some say they are tastiest sliced, cooked and served in a clear chicken soup.

• Chopped coarsely, then lightly blanched, dandelion leaves are excellent in a cheese omelette.

• Substitute dandelion greens for spinach in a variety of dishes, including eggs florentine, quiche and frittata.

• Use dandelion greens in a delicious linguine dish with a creamy goat cheese and prosciutto. To serve four, you will need about 3 L (12 cups) of dandelions, chopped and sautéed in olive oil with garlic, until the leaves wilt; then cook covered for 10 minutes in 250 mL (1 cup) of chicken stock. Add 150 g (6 oz) creamy goat cheese, crumbled and 125 g (4 oz) prosciutto, cut into thin strips. Mix in pot with al dente linguine. Season to taste with salt and pepper and serve.

Nutritional Value

Collards

(per 100 g = 3 1/2 oz)

			RDA**	%
calories*	(kcal)	40		
protein*	(g)	3.6		
carbohydrates*	(g)	7.2		
fat*	(g)	0.7		
fibre	(g)	0.9	30	3
vitamin A	(IU)	6500	3300	196.9
vitamin C	(mg)	92	60	153
thiamin	(mg)	0.2	0.8	25
riboflavin	(mg)	0.3	1.0	30
niacin	(mg)	1.7	14.4	11.8
sodium	(mg)	43	2500	1.7
potassium	(mg)	401	5000	8.02
calcium	(mg)	203	900	22.5
phosphorus	(mg)	63	900	7
magnesium	(mg)	57	250	22.8
iron	(mg)	1	14	7.1
water	(g)	86.9		

*RDA variable according to individual needs
**RDA recommended dietary allowance (average daily amount for a normal adult)

Curried collards with potatoes

(serves 4)

Ingredients

1-2	**large bunches of collards**	
5	**medium potatoes**	
45-60 mL	**butter or ghee (clarified butter)**	**3-4 TBSP**
2.5 mL	**whole black mustard seeds**	**1/2 tsp**
1	**medium onion, minced**	
2	**garlic cloves, minced**	
2 mL	**curry powder**	**1/4 tsp**
2.5 mL	**garam masala**	**1/2 tsp**
2.5 mL	**turmeric**	**1/2 tsp**
	salt to taste	

Instructions

• Clean, chop or shred collards, then make sure they are dry.

• Peel and cube potatoes. Boil them about 10 minutes, until just tender, but not too soft. Rinse in cold water, then drain well.

• In a heavy skillet, heat butter or ghee to medium and add mustard seeds. Stir, then add onion, garlic, curry, garam masala (a combination of cinnamon, nutmeg, cloves, etc.), turmeric and a little salt. Stir for a minute or two.

• Add the collards and potato, stir, then cover and cook for about 5 to 10 minutes on medium to low heat, stirring frequently.

• Serve as a vegetable side dish with chicken, fish or other Indian dishes.

Jerusalem artichoke

GOOD SOURCE OF IRON AND THIAMIN
LOW IN CALORIES AND SODIUM

Origin

• Also called sunchokes in the United States, Jerusalem artichokes are native to North America and were cultivated by Native North Americans long before the Europeans arrived. The first European of note to discover them in the early 17th century and introduce them to France, was the explorer Samuel de Champlain. In France they were named topinambours, after some members of a Brazilian tribe who were then visiting Paris — the difference between Canada and Brazil obviously meant nothing to the average European at that time!

• Another member of the Compositæ family, Jerusalem artichokes are in fact distantly related to artichokes, though this was probably not known to the early settlers, who thought they tasted like artichoke hearts. They are tubers of a type of sunflower, "girasol" (Spanish) or "girasole" (Italian), which literally means turning to the sun, a characteristic of all sunflowers. This was probably corrupted to the word "Jerusalem".

• Jerusalem artichokes quickly became a popular food in the colonies and were called sunroots. They later became widespread across Europe, especially in England and Italy, long before the potato was commonly used. (Potatoes were thought to be poisonous when first introduced to Europe.) As potatoes became accepted, Jerusalem artichokes began to be regarded as a food for the poor and fell out of favour.

• Along with salsify, Jerusalem artichokes should be eaten in moderation at first, as they produce gas in some people.

• Today, Jerusalem artichokes are grown in many parts of North and South America, Europe and Africa.

Varieties and description

• These small to large, irregularly shaped tubers have a very thin light brown skin, which is sometimes tinged with red or purple, depending on the soil in which they are grown. The creamy-white flesh is sweet, nutty, crunchy and juicy — somewhat reminiscent of water chestnut or jicama.

• Some varieties have been developed for more regularly shaped tubers.

Availability

• Available all year, Jerusalem artichokes are most plentiful from October to May. Those harvested in the summer are not as sweet as winter ones.

Selection

• Choose Jerusalem artichokes that are small, fresh, firm, dry and smooth-skinned. Avoid those tinged with green.

Storage

• It's best to keep Jerusalem artichokes in a plastic bag with a piece of paper towel to absorb excess moisture. They should be used within a week or two of purchase. Never freeze Jerusalem artichokes even after they are cooked or they will discolour.

Preparation

• Jerusalem artichokes are good raw or cooked. They should be well scrubbed, then peeled or not, as desired; the skin is very nutritious but not visually pleasing in some dishes. If they are to be used raw with a dip or in salad, they should be sprinkled with lemon juice to prevent them from darkening. If they are to be cooked, they can be peeled easily after cooking.

• The important thing about cooking Jerusalem artichokes is not to overcook them — even one or two minutes too long and they become mushy and lose their flavour. Boil or steam them with a squeeze of lemon juice for about 12 to 20 minutes (depending on their size) or until they are just tender, then serve them whole or mashed, with seasoning and butter. Use leftovers combined with minced onion to make hash-browns. If used in stews, add them for the last 20 minutes of cooking. Jerusalem artichokes are excellent pickled.

Suggestions

• Creamy baked Jerusalem artichokes are delicious: Simply steam or boil them until tender, cool, then remove the skins. Sauté a large onion with some fresh herbs in a little butter. Slice the Jerusalem artichokes and arrange them in a baking dish. Sprinkle on a little salt and pepper, then top with the onions, a generous layer of grated cheddar cheese, a thin layer of breadcrumbs, and a few dots of butter. Bake about 15 to 20 minutes in a moderate oven, then put the dish under the grill for a minute or two, until it browns on top.

• Jerusalem artichokes are delicious in puréed soups, especially when combined with leeks.

• They are crunchy yet juicy in a stir-fry.

• Sliced finely and marinated, grated in cole slaw or used in a salad, they are a good low-calorie, high-fibre, crispy addition.

• For an appetizing alternative to potatoes, try roasting Jerusalem artichokes with chicken or roast meats for the last 30 minutes or so of preparation. Turn them once or twice to get them nicely browned.

Jerusalem artichoke salad

(serves 4)

Ingredients		
4-6	Jerusalem artichokes	
1	green apple	
1	orange	
2	beets, cooked or raw	
2	stalks celery	
8	whole lettuce leaves	
1	carrot, grated	
	Dressing	
45 mL	olive oil	3 TBSP
30 mL	white wine vinegar	2 TBSP
5 mL	Dijon mustard	1 tsp
2.5 mL	Worcestershire sauce	1/2 tsp
	salt and pepper to taste	

Instructions

• Make the dressing first by mixing together all the ingredients.

• Peel and coarsely grate the Jerusalem artichokes, then toss them immediately with about 2/3 of the dressing, to prevent them from discolouring.

• Dice the apple. Segment and seed the orange, and cut it into small pieces. Slice the celery. Add the apple, orange and celery to the Jerusalem artichokes.

• Peel and coarsely grate the beets, then toss them with the remaining dressing. Arrange them in a ring on a bed of lettuce, then place the rest of the salad in the middle. (If you mix it all together, everything will turn pink from the beets.)

• Sprinkle the grated carrot over the top.

Jerusalem artichoke salad →

Pickled Jerusalem artichokes

Ingredients		
500 g	Jerusalem artichokes	1 lb
	juice of 1/2 lemon	
250 mL	pearl onions	1 cup
6	whole garlic cloves	
2	red bell peppers	
125 mL	white wine vinegar	1/2 cup
250 mL	water	1 cup
	zest from 1/2 lemon	
60 mL	sugar	1/4 cup
30-45 mL	coarse salt	2-3 TBSP
15 mL	celery salt	1 TBSP
5 mL	turmeric	1 tsp
30 mL	pickling spice	2 TBSP
5 mL	dry mustard	1 tsp
1	large bunch fresh dill	
10 mL	capers (optional)	2 tsp
1-2	whole chili peppers (optional)	

Instructions

- Peel the Jerusalem artichokes and slice them in rounds. Keep them in a bowl of cold water with the juice of 1/2 a lemon until you are ready.
- Peel pearl onions and garlic, leaving them whole. Cut the red bell peppers in strips.
- Heat the vinegar with a cup of water, and add lemon zest, sugar, salt, celery salt, turmeric, pickling spice and dry mustard. Stir until sugar is dissolved.
- Remove the Jerusalem artichoke slices from the lemon water, reserving the liquid.
- In a gallon pickling jar, make layers with whole sprigs of fresh dill, Jerusalem artichoke slices, peppers and pearl onions, sprinkling capers, garlic bulbs and the chili peppers between the layers. (Cauliflower is another good optional ingredient.)
- Pour the hot vinegar mixture over the vegetables, then add enough of the lemon water from the Jerusalem artichokes to just cover the vegetables.
- Put a small plate with a weight (such as a clean stone) on it to keep the vegetables immersed.
- Keep in a cool, dark, ventilated place for 2 to 3 weeks before refrigerating.

Nutritional Value
Jerusalem artichoke
(per 100 g = 3 1/2 oz)

			RDA**	%
calories*	(kcal)	7-75***		
protein*	(g)	2.3		
carbohydrates*	(g)	16.7		
fat*	(g)	0.1		
fibre	(g)	0.8	30	2.7
vitamin A	(IU)	20	3300	0.6
vitamin C	(mg)	4	60	6.6
thiamin	(mg)	0.2	0.8	25
riboflavin	(mg)	0.06	1.0	6
niacin	(mg)	1.3	14.4	9.1
sodium	(mg)		2500	
potassium	(mg)		5000	
calcium	(mg)	14	900	1.6
phosphorus	(mg)	78	900	8.6
magnesium	(mg)	11	250	4.4
iron	(mg)	3.4	14	24.3
water	(g)	79.8		

*RDA variable according to individual needs
**RDA recommended dietary allowance (average daily amount for a normal adult)
***Kcal range from 7 kcal/100 g for freshly harvested to 75 kcal/100 g after long storage

Horseradish

Origin

• Horseradish originated in the vast region from Eastern Europe to Western Asia. A member of the Crucifer family, it is related to numerous plants, including cabbage, broccoli, cauliflower, radish, arugula and mustard cress. It is not related to the so-called Japanese horseradish, wasabi.

• Horseradish is one of the bitter herbs mentioned in the Exodus story in the Old Testament. Known to all the ancient European civilizations, it was considered most valuable as a medicine for a whole range of ills, from coughs to gout. By the end of the Middle Ages it had spread across Europe, first gaining popularity in Germany and Central Europe and later in Scandinavia and England, where roast beef with horseradish sauce has long been a favourite.

• Horseradish now grows abundantly in North America and in many other parts of the world.

Description

• Horseradish root is light-to-yellowish brown, coarse-skinned, white-fleshed and very hard. It averages 30 centimetres (1 foot) in length.

• When horseradish is scraped or grated, a volatile oil is released, producing a characteristic sharp, pungent aroma that brings tears to the eyes.

Availability

• Available year round, fresh horseradish is sometimes hard to find, as the public is used to buying it already processed. It is most abundant in the spring and late fall.

Selection

• Choose firm, hard roots with no soft spots.

Storage

• Wrap the root in a slightly damp paper towel, place it in a paper bag and refrigerate in the crisper for up to a few weeks. If the root looks like it's beginning to shrivel or go soft in spots, prepare it immediately. Horseradish sauce can be kept for quite a while, but will lose strength. The same will happen if the grated root or sauce is frozen.

Preparation

• Wash horseradish thoroughly and peel off the skin. If any green flesh is present just under the skin, discard it as well, as it tastes bitter. Some people also discard the central core of a large root if it appears hard and woody.

• A small amount of the root can be grated by hand over a variety of dishes just before serving. A large amount is most easily prepared in a food processor or a blender. Chop it in small pieces before processing. Its white flesh will turn brown when exposed to the air unless lemon juice or vinegar is added immediately.

• When cooked, the pungent flavour disappears and the taste resembles turnip.

Suggestions

• Fresh grated horseradish is wonderful in potato, tuna, avocado and chopped egg salads, in vinaigrettes, and simply sprinkled over a variety of soups (especially creamed soups), cooked vegetables and meats. For a great taste, try a generous amount in your hamburger mixture.

• Grated horseradish is also delicious in sandwiches.

• Add grated red or yellow beetroot to give horseradish sauce an attractive colour. Add grated turnip and a pinch of sugar or some grated apple for sweetness, if desired. Add a touch of mustard or garlic for additional pungency.

• Horseradish combines well with cream, sour cream, yogurt or mayonnaise to make sauces. It is excellent in a tomato-based sauce made to accompany crab, shrimp or seafood cocktails.

Smoked trout with horseradish sauce

(serves 4)

Ingredients		
250 mL	grated fresh horseradish	1 cup
125 mL	white wine vinegar	1/2 cup
45-60 mL	mayonnaise	3-4 TBSP
	pinch of salt	
4	fillets of smoked trout	
	4 medium or 8 small tomatoes	
4	whole lettuce leaves	
1	cucumber	
	sweet coloured peppers	

Instructions

- Mix the first three ingredients thoroughly and season to taste.
- Slice off the top of each tomato and scoop out the centres; stuff the tomato shells with the horseradish sauce.
- Make four individual plates, or 1 large platter, decoratively arranging the fish on lettuce, the stuffed tomatoes, sticks of cucumber, and slivers or rounds of sweet coloured peppers.
- Keep the rest of the horseradish sauce refrigerated or frozen for another occasion.

Swiss chard

↑ Swiss chard

EXCELLENT SOURCE OF VITAMINS A AND C
EXCELLENT SOURCE OF IRON, MAGNESIUM AND OTHER MINERALS

Origin

• Swiss chard, sometimes called chard or leaf beet, is a hardy type of beet prized for its succulent stalks and leaves rather than for its slender, woody roots. Despite its name it has no special association with Switzerland. It originated in the Mediterranean region and was known to have spread as far as the Middle East and Persia more than 2,000 years ago. This vegetable, cultivated for several thousand years, is thought to be the ancestor of beetroot. Well documented by the ancient Greeks and Romans, who used it medicinally, it was later written about in every important Medieval herbal.

• Swiss chard is a member of the Chenopod family (literally the goosefoot family — so named for the shape of their leaves), along with beetroot, spinach and spinach beets.

• Today, Swiss chard is grown in every European country (especially France), some Asian and African countries, South America and, to a lesser extent, in North America.

Varieties and description

• Swiss chard is a prolific biennial that can be harvested several times each growing season. It has large, flat or crinkly, light to dark green leaves growing from smooth, broad, thick, central stalks or ribs that are usually white to pale green, or an attractive red in the case of rhubarb chard. Some varieties of rhubarb chard have dark red leaves as well.

• Both the leaves and stems are eaten, which leads some cooks to call it two vegetables in one. The taste of the leafy part is spinach-like, while the stalks have a fresh, delicate, crispy texture and flavour.

Availability

• Available from April to December, Swiss chard is most plentiful from June to October.

Selection

• The leaves and stalks of Swiss chard should be crisp and fresh-looking. Avoid wilted or brown leaves.

Storage

• Keep Swiss chard refrigerated, unwashed, in a perforated plastic bag, for no more than three or four days. It tastes best when fresh.

Preparation

• Discard a thin slice from the base of the stalk and wash the leaves thoroughly in a sink full of cold water. Swiss chard has more flavour when it is steamed or sautéed rather than boiled. It shouldn't be cooked in aluminum or iron pots or pans, which discolour it. Slightly tougher than spinach, it requires just a little more cooking. Many cooks separate the stalks from the leaves if they are wider than 1.5 centimetres (1/2 inch) and either cook them separately or add the leaves when the stalks are nearly done. The leaves, unlike the stalks, freeze well when they have been cooked.

• Treat the leaves like spinach and steam them whole or chopped until wilted. If they are very young and fresh, they can be eaten raw in salads. The stems are great steamed, whole or cut on the diagonal, and served like asparagus, hot or cold, in a variety of sauces, including vinaigrette, mornay and hollandaise. They require only a few minutes of steaming and should still be slightly crisp when done. They are a good substitute for bok choy (a Chinese green) in stir-fry recipes and they are excellent chopped and added to a soup a few minutes before it is finished cooking.

Suggestions

- Creamed Swiss chard soup is delicious. Swiss chard can be added to a great many soups, such as chicken, minestrone, lentil and vegetable.

- Use seasoned Swiss chard and ricotta cheese filling to stuff pasta such as cannelloni or tortellini.

- For a great vegetable dish, prepare a bed of steamed Swiss chard leaves in a baking dish, place the cooked stalks over them, mix crab meat into mornay sauce and pour it over the top. Add a sprinkling of Parmesan cheese, then breadcrumbs. Heat through in the oven, then brown the top. For a more elegant look, prepare in coquille St. Jacques shells or individual ovenproof dishes.

- The leaves are excellent for stuffing. Cut the stems from the leaves, then wilt the leaves by steaming or blanching. The stalks can be cooked, then minced and added to the stuffing or served as a side vegetable. Make your stuffing with minced meat or any grain sautéed with onions, garlic, seasoning and herbs, or the stuffing of your choice. Put a spoonful or two in the centre of each leaf, wrap it and place in a shallow baking dish with the loose end face down. Add tomato sauce, creamy cheese sauce or another braising liquid to barely cover the stuffed leaves, sprinkle Parmesan cheese over the top and bake in a moderate oven for about 45 minutes.

- In Europe, Swiss chard is often served hot or cold after being sautéed in a little olive oil, garlic, nutmeg, pepper and salt, with a generous squeeze of lemon or wine vinegar as a final touch. Sautéed shallots or onions and anchovies go well in this dish when served hot.

- For a delectable treat, try Jugoslavian chard pancakes: Cook a bunch of chard. Mince it finely, squeeze out excess water and put it in a bowl with grated cheese, a few spoons of sour cream and enough breadcrumbs to make it stick together well. Add 15 mL (1 TBSP) vegeta (a European mixed vegetable spice) or powdered chicken stock and a little salt and pepper. Mix well and form into small patties. (Exact measurements aren't necessary.) Dip each patty first into a plate with flour, then into a shallow bowl with a beaten egg thinned with a little water and, finally, into a plate with breadcrumbs. Fry the patties in a Teflon pan with just a little oil until they are golden brown. Serve with sour cream or yogurt. Any extras will freeze well.

Nutritional Value
Swiss chard
(per 100 g = 3½ oz)

			RDA**	%
calories*	(kcal)	25		
protein*	(g)	2.4		
carbohydrates*	(g)	4.6		
fat*	(g)	0.3		
fibre	(g)	0.8	30	2.7
vitamin A	(IU)	6500	3300	196.9
vitamin C	(mg)	32	60	53.3
thiamin	(mg)	0.06	0.8	7.5
riboflavin	(mg)	0.17	1.0	17.0
niacin	(mg)	0.5	14.4	3.5
sodium	(mg)	147	2500	5.9
potassium	(mg)	550	5000	11.0
calcium	(mg)	88	900	4.3
phosphorus	(mg)	39	900	4.3
magnesium	(mg)	65	250	26.0
iron	(mg)	3.2	14	22.9
water	(g)	91		

*RDA variable according to individual needs
**RDA recommended dietary allowance (average daily amount for a normal adult)

Rhubarb chard ↓

Chili peppers

↑ Finger hot

Origin

• Native to the New World, probably Brazil, chili peppers have been cultivated for more than 7,000 years. Used extensively by the Incas and later by the Aztecs, they gradually spread south, as well as north to the American Southwest, where the Pueblo Indians lived. They were discovered in the tropics by early European explorers and brought back to Spain by the beginning of the 16th century. The Portuguese introduced them to Africa, and later to India in the early 17th century, where they soon became a very important spice, essential to Asian cuisine.

• Chili peppers, very closely related to sweet peppers, belong to the Solanaceæ family, along with tomatoes, potatoes, eggplant and many others. The main difference between sweet and hot peppers is the latter's high content of capsaicin, a powerful, volatile oil that can burn the eyes and mouth if handled carelessly. Curiously, this oil provides the most effective relief in the treatment of shingles.

• Chili is used as a spice, rather than as a vegetable. It is extremely popular all over the world, particularly in China (Szechuan and Hunan provinces especially), Southeast Asia, India, Africa, some Mediterranean countries, Mexico and many Latin American countries, as well as in Creole cooking. Chilis are also grown in large quantities in California, where Mexican food is popular.

Varieties and description

• There are numerous varieties of chilis, which cross breed easily, making them difficult to distinguish. From mildly hot to blistering hot, chilis can vary greatly from one variety to another, between plants of the same variety, and from one chile to another on the same plant.

• All chili peppers start off green and some varieties are left on the bush to ripen to red, yellow, orange, dark purple or dark brown. They vary both in shape and size, and generally the smallest ones are the hottest. Some of the best known varieties are cayenne, jalapeño, serrano, Anaheim, poblano, pasilla, Mexi-bell, Fresno, hot banana, habanero, chilaca, mulato, Santa Fe grande, fushimi and Tabasco.

• Long green or red Anaheims are among the most common and mildest, along with the short, wide, darker green poblano or ancho and pasilla peppers. Also mild is the Mexi-bell, a hybrid that resembles a small sweet green bell pepper, the large yellow hot banana (not to be confused with the large sweet banana pepper) and the green Japanese fushimi peppers.

• Medium to hot chilis include the plump, light green or red Fresnos, the yellow, orange or red Santa Fe grandes, the long, thin, nearly black chilaca and the dark green, red or orange plump jalapeños, which are the hottest in this category.

• Among the hottest chilis are the tiny, thin, fiery-hot red and green cayennes, also known as Thai peppers, the short, chunky, green, yellow or red habaneros and the small green or red serranos.

1. **Alma (hot apple)**	2. **Poblano**	3. **Crimson Hot (Hybrid)**	4. **Red Anaheim**	5. **Anaheim**
6. **Hot banana**	7. **Fresno**	8. **Jalapeño**	9. **Golden cayenne**	10. **Serrano**
11. **Torito**	12. **Big cayenne**	13. **Hot cherry**	14. **Finger hot**	15. **Long hot**
16. **Habanero**				

Availability

- Available all year, with the peak season in the summer.

Selection

- Chilis should be glossy, with fresh-looking skins.

Storage

- Fresh chilis can be kept refrigerated in a paper bag for up to a week or more. The green ones usually keep longer. Red chilis can be dried for later use and all chilis can be pickled.

- For freezing, remove skins and veins, then either blanch or broil.

↓ Marinated peppers

Preparation

- Use chili in moderation at first; add a small quantity of finely minced chili or add whole chilis, which can be removed when you think the food is hot enough. Most of the capsaicin, which causes the hotness, is in the veins or inner ribs and to a lesser extent in the seeds. Halve the chili and remove the veins and seeds before use.

- Chilis can be soaked in cold water, with or without a little vinegar, to draw off some of their heat. Roasting them under the broiler until they brown gives them a good flavour and makes removing the skins easy; put them in a paper or plastic bag for about 10 minutes afterward, to let the steam get under their skin, then peel it off.

- Careful handling of chilis is essential; people with sensitive skin should wear gloves. Wash hands, knife and chopping board thoroughly with soap and water when you are finished. Never touch your eyes or a cut while preparing chilis. Some capsaicin may linger for a while even after you've washed your hands, as it is not water soluble. Remedies for burning mouths include cooked rice, milk, ice cream, yogurt and sugar or sweets.

Suggestions

- For a delicious brunch, try Mexican huevos rancheros: In a large frying pan, sauté minced onion with a little minced chili, then add a diced sweet pepper, slivers of ham, and any leftover firm vegetables, such as sweet corn, peas or sliced beans. Finally, add some diced tomato and stir for a minute. Beat as many eggs as required in a bowl, then add them to the frying pan, stirring until scrambled. Season with salt. Serve with fresh bread, refried beans and guacamole or slices of avocado sprinkled with lemon juice.

- Fresh chilis are great in soups, stir-fries, stews (especially chili con carne) and sauces. They are essential for Indian cuisine.

- A little chili will perk up bland bean or vegetable dishes and some pasta dishes.

- A favourite trick in Louisiana and the Southwest is to add a little finely minced chili to bread dough, cornbread batter or such dishes as cheese soufflé.

- Chili is good in marinades, barbecue sauces, relishes and all sorts of condiments.

Mushrooms

(Chanterelles / morels / porcini)

Origin

• Of more than 2,000 edible species of mushrooms, less than two dozen are available commercially. Of those, some, such as chanterelles, morels and porcini, cannot be cultivated and are still gathered from the wild. Though mushrooms have been known to man for thousands of years, their origins are not certain, as the same varieties are found in many parts of the world, probably because their tiny spores can be carried vast distances by the wind.

• Chanterelles, or pfifferlinge, are extremely popular in Europe. Morels are one of the most rare and sought-after of all the wild mushrooms. Porcini (por chee nee), so-named because it is a favourite mushroom of pigs, is the Italian name for Boletus edulis. Called ceps or cèpes in France and steinpilze (stone mushroom) or herrenpilze (master mushroom) in Germany, they are sometimes called king bolete.

• Wild mushrooms are considered to be more like meat than any other vegetable, as they contain a lot of protein, minerals and vitamins, few carbohydrates and no sugars. They are known to be excellent for the digestion.

• These three mushrooms grow in the wild in many European countries, as well as in some parts of the western and eastern United States and Canada.

Okra

VERY LOW IN CALORIES
GOOD SOURCE OF VITAMIN C
SOURCE OF CALCIUM AND
OTHER MINERALS

Origin

• Known also as gumbo or lady fingers (because of its size and shape), okra is a member of the mallow family. There are several medicinally useful varieties of mallow. The most economically important members of mallow are the various species of cotton.

• Okra, a beautiful ornamental plant, is closely related to hibiscus and to Jamaica sorrel, a flower used widely in Mexico, the Caribbean and parts of Africa to make delicious drinks and jams.

• Okra originated in Africa or Asia and has been cultivated for several thousand years. It was brought to North and South America by African slaves. Angolans called the vegetable "ngombo", which became the Creole word gumbo. At first it meant any stewed dish containing okra; later it came to mean any Creole stew. Possibly because it was a favourite food of the slaves, it became known as a food for the poor, and therefore was ignored by many.

• Okra has always been important in the American South, as well as in Latin American, African, Mediterranean, Middle Eastern, Indian and Southeast Asian cooking. It is just beginning to be appreciated in the rest of North America.

Varieties and description

- Okra must be picked and eaten while still unripe, as it becomes too coarse and fibrous when fully developed. This prolific plant is harvested every day or two so the pods don't mature. As long as the pods are picked often, the plant keeps producing more.

- Though these slender bright green edible pods, topped by small pointed caps, can grow up to 22 to 25 centimetres (9 or 10 inches) long, the smaller ones of 5 to 10 centimetres (2 to 4 inches) taste best. Its skin, smooth in some varieties and fuzzy in others, has shallow grooves running lengthwise. Its somewhat sticky flesh, which makes okra a great natural thickener for soups and stews, contains numerous soft, edible seeds. Okra has an unusual texture and a mild, agreeable flavour.

Availability

- Available all year, okra's peak season is from June to October.

Selection

- Choose tender though not limp, fresh-looking okra pods. If they are too old they become fibrous and have a gummy texture.

Storage

- Okra is best when used within a day or two of purchase. Keep it either in a paper bag or covered by a paper towel in a plastic bag on the top shelf of your refrigerator.

Preparation

- If necessary, rub off the fuzz with a paper towel or a soft vegetable brush. Wash and drain.

- If cooking it whole, discard the stems and tips of the caps, without cutting into the pod. If the okra is longer than 5 to 8 centimetres (2 to 3 inches), whittle off some of each cap. Steam or boil until just tender but not overcooked (about 10 minutes), or simmer longer in a casserole, stew or mixed vegetable dish. Once cooked, okra can be left whole or sliced, according to the recipe.

- For soups and certain other dishes, cut off the cap altogether, slice the okra in rounds and cook, according to directions. This releases some of its sticky juices, which act as a thickener.

- Okra is very good and possibly less sticky when prepared in a microwave.

- Okra can be blanched, then frozen.

- Please note that iron, copper and aluminum pots or pans will discolour okra, though not affect its taste.

Suggestions

- Okra is good steamed, boiled, fried in batter, sautéed, stewed or pickled, but not puréed. It is used in a wide variety of Creole dishes, usually accompanied by rice.

- Certain vegetables — such as tomatoes, eggplant, onions, peppers and sweet corn — combine wonderfully with okra, as do many flavourings, including cilantro, thyme, oregano, lemon and vinegar.

- In the American South okra is often eaten with crispy bacon crumbled on top; in the Middle East, in lamb stews.

- Steamed or boiled until tender, okra is appetizing served cold with vinaigrette or hot with lemon, seasoning and melted butter or in a tomato-based sauce.

- Okra is excellent in many Indian dishes. For a simple preparation, stir-fry whole okra with finely chopped onion, curry powder, garlic, ginger, a pinch of chili, salt, finely chopped cilantro and possibly a little grated coconut. After a few minutes, add a little tomato paste and enough water to keep everything moist. Cover and cook until tender, from 10 to 20 minutes, depending on the size of the pods.

- To thicken a hearty soup or stew, slice okra in small rounds and add it for the last 10 to 12 minutes of cooking. Okra is especially good in fish or seafood soups or stews.

- Leftover okra is great sliced and added to Spanish omelettes or scrambled eggs.

- Cube 450 g (1 pound) lamb and brown chunks on all sides in a large uncovered saucepan on low heat for about an hour. Do not add any fat unless the meat is very dry. Add one small finely minced onion, two cloves of finely chopped garlic, 450 g (1 pound) okra cut in 2 centimetres (1 inch) lengths and a little salt. Cover and cook for 10 minutes, then stir in 250 mL (1 cup) tomato sauce, 30 mL (2 TBSP) lime juice and enough water to barely cover the ingredients. Season to taste, then cover and simmer about 20 minutes or until tender. Serve with rice.

Okra ratatouille

(serves 4)

Ingredients		
1	large onion, diced	
1	garlic clove	
30-45 mL	olive oil	2-3 TBSP
250 g	okra	1/2 lb
1/2	red pepper, slivered	
1/2	yellow pepper, slivered	
4-6	plum tomatoes, sliced	
1	bay leaf	
5 mL	fresh thyme	1 tsp
30 mL	sherry or white wine	2 TBSP
	pinch of chili (optional)	
	pinch of sugar	
	salt and pepper to taste	

Nutritional Value

Okra

(per 100 g = $3^{1}/_{2}$ oz)

			RDA**	%
calories*	(kcal)	29		
protein*	(g)	2		
carbohydrates*	(g)	6		
fibre	(g)	1.0	30	3.3
vitamin A	(IU)	490	3300	14.8
vitamin C	(mg)	20	60	33.3
thiamin	(mg)	0.13	0.8	16.3
riboflavin	(mg)	0.18	1.0	18
niacin	(mg)	0.9	14.4	6.3
sodium	(mg)	2	2500	0.08
potassium	(mg)	174	5000	3.5
calcium	(mg)	92	900	10.2
phosphorus	(mg)	41	900	4.6
magnesium	(mg)		250	
iron	(mg)	0.5	14	3.6
water	(g)	91.1		

*RDA variable according to individual needs
**RDA recommended dietary allowance (average daily amount for a normal adult)

Instructions

- Sauté the onion and garlic in olive oil for 2 to 3 minutes at medium heat, until the onion becomes translucent.
- Prepare the okra, then add it whole to the frying pan with all the other ingredients.
- Cover and simmer until the okra is tender, about 20 to 25 minutes, taking care not to overcook it. You may have to add small amounts of water to keep it moist as it cooks.
- Serve as a vegetable side dish with lamb, beef, chicken or fish and rice.

Dry shallots and pearl onions

VERY LOW IN CALORIES AND SODIUM
SOURCE OF IRON AND VITAMIN C

Origin

• The aristocratic shallot and the tasty pearl onion are members of the Lily family along with more than 300 species of onion-like vegetables that include leeks, scallions, garlic and numerous species of onions. Known to prehistoric man, the onion's true origins are unclear, though most species are thought to be native to Western and Central Asia.

• The ancient Greeks grew shallots and traded them to other countries. They were highly prized by the Romans, both as a food and an aphrodisiac. Some believe that the word shallot is derived from a Middle-Eastern city called Ashkelon or Ascalon and that they were introduced into England and France during the 13th century by crusaders returning home. However, it has never been proved.

• Today, the top quality dry shallots are grown in France, in the Brittany region. They are also grown in other European countries and in Canada and the United States.

• Pearl onions may have originated in Southern Europe or the Middle East. Long popular in Europe, pearl onions were introduced in the early part of this century by immigrants to California, where they are now grown in large quantities.

Shallots ↑

Varieties and description

• A shallot is a small bulb, roughly the size of a small head of garlic, that sometimes separates into 2 or 3 cloves under its skin. The most widely cultivated ones are pink shallots, which have a papery, pinkish skin and an oblong-shape, and copper-yellow shallots, which have a papery, yellowish skin and are elongated. Both are graded in small, medium and jumbo sizes.

• The tiny, marble-sized, mild pearl onion, sometimes called a cocktail onion, comes in three colours — silvery white, red and golden yellow.

Availability

• Thanks to the benefits of cold storage, shallots and pearl onions are available year round.

Selection

• Shallots and pearl onions should be firm, with dry, papery skins. Avoid those that have started to sprout green shoots and those with dark spots.

↑ Pearl onions

Storage

• Keep shallots and pearl onions in a cool, dry, dark place and they will stay in excellent condition for about a month. Or keep them refrigerated and use within a couple of weeks. Once they are peeled or cut up, they can be wrapped in plastic or placed in a jar and covered with olive oil, then refrigerated. Be sure to use this aromatic olive oil in salad dressings afterward.

Preparation

• Cut both ends off each shallot, slit lengthwise and the peel should come off easily. Slice thinly, mince or leave whole, according to the recipe.

• Keep the outer peels of shallots to use in soup stocks; they add a good colour and a great flavour to soup. Discard peel when the stock is cooked.

• The easiest way to peel pearl onions is to blanch them in boiling water for a minute, then plunge them into cold water. Cut a thin slice off the root end, hold them by the stem end and pop them out of their skins.

Suggestions

• Mince shallots finely and add them to a vinaigrette to dress a salad or an avocado for a distinctive gourmet taste.

• Shallots are delicious when roasted whole with very small turnips and chunks of potato, flavoured with rosemary, thyme, salt, pepper and a little olive oil.

• For a hearty, healthy soup, sauté 8 to 10 shallots with 2 cloves of garlic in a little olive oil. Add 1 or 2 grated carrots, 250 mL (1cup) orange or brown lentils, 625 mL (2 1/2 cups) water and 1 whole chili. Boil, simmer until the lentils are cooked, then discard the chili. Add a handful of finely chopped cilantro, a squeeze of lemon and season to taste.

• Shallots add a great flavour to soups, coq au vin, casseroles, pasta dishes or stew. Pearl onions can be used for the same purpose but give a much milder flavour.

• Minced and sautéed with leftover boiled potatoes and seasoning, shallots give a whole new flavour to hash browns.

• Shallots are essential for several traditional French sauces, such as beurre blanc, sauce marchand de vin and sauce à la bordelaise.

• Pearl onions are good whole or sliced in salads. They are excellent pickled and used in sauces.

• Thread a few skewers with pearl onions and cook them on the barbecue or under the grill along with your main dish.

Nutritional Value

Dry shallots

(per 100 g = 3 1/2 oz)

			RDA**	%
calories*	(kcal)	72		
protein*	(g)	2.4		
carbohydrates*	(g)	16.8		
fat*	(g)	0.2		
fibre	(g)	0.8	30	2.7
vitamin A	(IU)	3300		
vitamin C	(mg)	8	60	13.3
thiamin	(mg)	0.06	0.8	7.5
riboflavin	(mg)	0.02	1.0	2.0
niacin	(mg)	0.2	14.4	1.4
sodium	(mg)	12	2500	0.5
potassium	(mg)	334	5000	6.7
calcium	(mg)	36	900	4.0
phosphorus	(mg)	60	900	6.6
magnesium	(mg)		250	
iron	(mg)	1.2	14	8.6
water	(g)	69.2		

*RDA variable according to individual needs
**RDA recommended dietary allowance (average daily amount for a normal adult)

Pearl onion sauce

(serves 4)

Ingredients		
30 mL	olive oil	2 TBSP
15 mL	butter	1 TBSP
1	basket pearl onions, peeled	
4	dry shallots, minced	
1	garlic clove, minced	
	pinch of nutmeg	
8	chanterelle mushrooms	
30 mL	fresh dill, finely chopped	2 TBSP
30 mL	flour	2 TBSP
60 mL	dry white wine or sherry	1/4 cup
375 mL	chicken or fish stock	1 1/2 cups
	salt and pepper to taste	
30 mL	parsley, finely chopped	2 TBSP

Instructions

- Heat the oil and butter in a large skillet that has a lid.
- Add pearl onions, shallots, garlic and nutmeg and sauté for a few minutes on medium heat, stirring frequently.
- Slice and add the mushrooms, then add the dill. Stir for another minute or two, then sprinkle in the flour and mix well.
- Add the wine, mix until smooth, then slowly add the stock, stirring continuously until it begins to thicken.
- Cover and simmer for about 5 more minutes, or until the onions are done, stirring occasionally. Adjust seasoning.
- Serve with grilled fish and baked potatoes or boiled new potatoes, garnishing with parsley.

Confit of shallots

(serves 4)

Ingredients		
500 g	whole shallots, peeled	1 lb
30-45 mL	butter	2-3 TBSP
30-45 mL	sugar	2-3 TBSP
	salt and pepper to taste	
125 mL	water	1/2 cup

Instructions

- To cut down cooking time, cover shallots with water, heat to boil, then simmer for 5 to 10 minutes until softened. Drain.
- Melt butter in a frying pan, add shallots and stir to coat them thoroughly.
- Cook on a medium heat for a few minutes, then sprinkle on sugar, salt and pepper. Stir for a minute, then add the water. Bring to the boiling point, then turn the heat to low.
- Stir from time to time until the shallots are browned and caramelized and only a little syrupy liquid is left.
- For a variation of this delicious vegetable side dish, substitute 15 mL (1 TBSP) of red wine vinegar for an equivalent amount of water.

↓ Confit of shallots

Squash

Orange-fleshed squash is an excellent source of vitamin A
Low in calories and sodium
Good source of fibre

Origin

• Squash, or edible gourds, and melons are members of the Cucumber family. Indigenous to North, Central and South America, they were a staple food of pre-Columbians, along with corn and beans, for more than 8,000 years before the first European explorers. Today, they are grown throughout the world.

• The word squash derives from the Algonquin word "askutasquash", which means "eaten raw" and probably referred to a kind of a summer squash encountered by early settlers.

• Calabaza, which means gourd in Spanish, now applies to a particular kind of squash still very popular in Latin America and the Caribbean.

• Kabocha means gourd in Japanese and is used for a few varieties of tasty squash that have been improved in Japan since World War II and are now grown in North America.

Varieties and description

• Squash are usually divided into summer and winter categories, which do not refer to their seasonal availability.

• Summer squash, often small, are immature, soft-skinned and perishable, with a very crunchy texture and a high water content. They are 100 percent edible and always have white flesh.

• Winter squash, usually larger, are fully mature, hard-skinned and have a long shelf life. They generally have a sweeter, nuttier taste and most are orange-fleshed.

• The very small, soft seeds of summer squash are usually eaten along with the flesh. The larger, harder seeds of the winter squash are discarded, or they can be washed, salted if desired, baked and enjoyed as a snack.

• Summer squash includes the smooth-skinned, long or round, green or yellow **zucchini;** white, pale or dark green, pear-shaped **chayote;** green and yellow striped, zucchini-shaped **cocozelle;** green or yellow, smooth or bumpy-skinned **crookneck** and **straight neck squash;** and flying saucer-shaped, scallop-edged, white, yellow, green or striped **patty pan,** cymling or **scallop squash.**

• Winter squash includes the popular fluted **acorn squash,** with its dark green or orange skin and orange flesh; the smooth, pear-shaped **butternut,** with its cream-coloured skin and orange flesh; the smooth, turban-shaped **buttercup,** which is green or orange, with a paler "cap", and has yellow flesh; the large, cylindrical **banana squash,** whose shell is ivory to pink; the club-shaped **delicata,** which is green and beige-striped and has yellow flesh; the pumpkin-shaped **kabocha** or Japanese squash, with its light grayish-green skin and bright yellow-to-orange flesh; the very small, pumpkin-shaped **sweet dumpling squash,** which is green and white streaked and has yellowish-orange flesh; the very small, round and ridged golden nugget squash, whose shell and flesh are both orange; the ornamental **turban squash,** whose bumpy skin is green streaked with various colours; the unusual **spaghetti squash,** which is round to oval and cream or yellow-shelled; and, of course, the beautiful and familiar orange **pumpkin.**

Availability

• Some summer squash, especially zucchini, is available year round, with peak supplies from April to September.

• Winter squash is also available year round, though at times some varieties are hard to find during the summer. The peak season is from September to March.

Selection

• All squash should have a solid, heavy feel: Squash that feels light for its size may be dehydrated.

• Summer squash should have firm though tender, sleek, unblemished skin. A shiny skin indicates it has been picked when young and tasty. Small to medium-sized are best.

• Winter squash varies widely in size. Some of the largest, like hubbard, calabaza and banana, often are cut and sold in chunks. Make sure the stem is still attached to any whole winter squash. The hard shells of winter squash should be undamaged and have a dull, not shiny skin, which indicates that it has been picked when fully mature.

1
2
3
3
2
4
1
4
1
1. Kabocha
2. Butternut
3. Spaghetti
4. Acorn squash

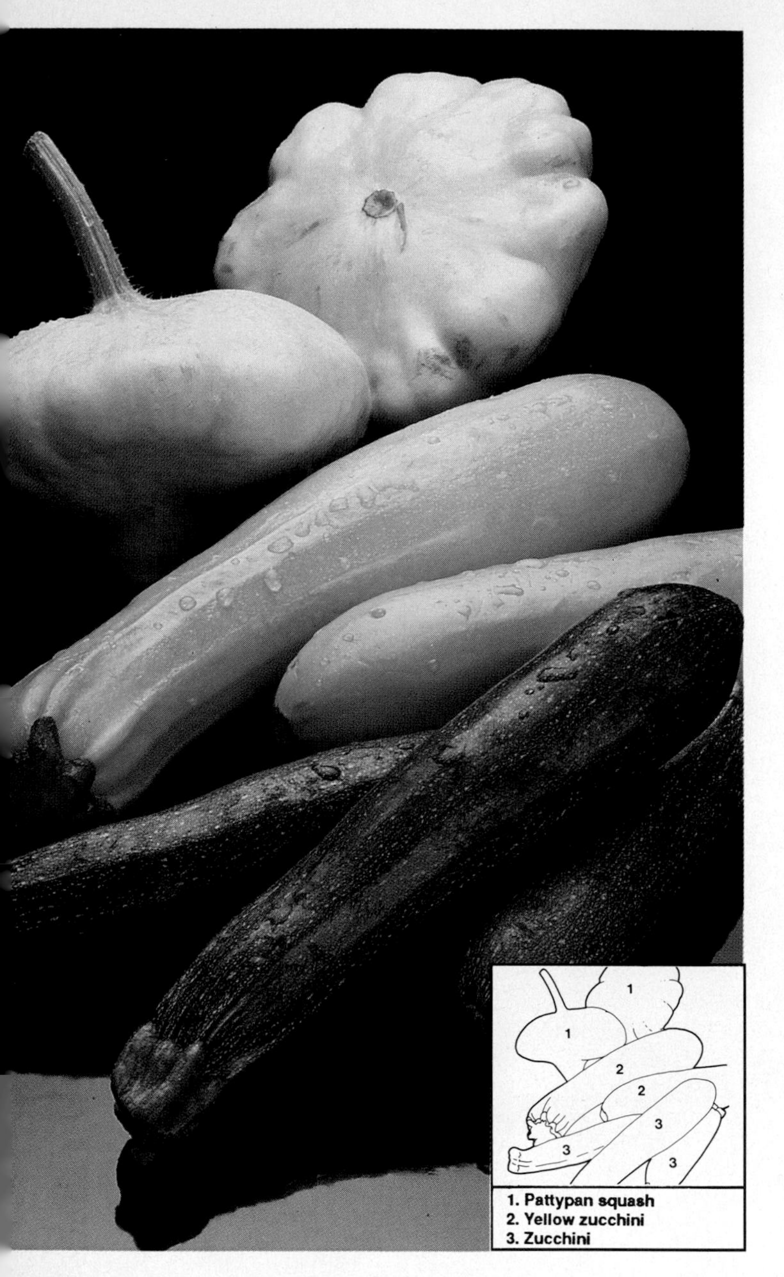

1. Pattypan squash
2. Yellow zucchini
3. Zucchini

➠ Storage

- Keep summer squash refrigerated in an unsealed plastic bag and use within 3 or 4 days of purchase. Handle carefully, as the soft skin is easily nicked.

- Never refrigerate winter squash unless it has been cut open, then wrap it in plastic and use within a day or two. It can be kept for a number of weeks (some varieties will keep for several months) stored somewhere dry and cool, not cold, out of direct sunlight. Smaller squash has a shorter shelf life and is best used within a few weeks of purchase.

➠ Preparation

- Summer squash has a high water content. Be careful not to overcook it or it will turn out mushy instead of crisp and delicious. It is best steamed or sautéed (whole if very small or cut in rounds or sticks), though it is also good lightly boiled, baked or grilled. If boiling, use a very small amount of water — cooking time will be only a matter of minutes, depending on the size of the pieces.

- Wash summer squash and discard a thin slice from each end. Zucchini, chayote, cocozelle and crooknecks never need peeling, but some of the others can be peeled if desired.

- Patty pans are good stuffed and baked; after blanching, cut off the stem, carefully dig out the seed cavity and some of the flesh, and stuff with breadcrumbs, minced onion, herbs and grated cheese, or sweet corn, diced red peppers and minced scallions. Bake about 20 minutes, or until tender.

- Winter squash should be cut open very carefully, as the knife can easily slip off the hard shell. Use a hammer or mallet starting near the knife handle and tap it in gently until the squash splits. The seeds can be removed before or after cooking.

- Small winter squash is best when baked until tender. Large winter squash can be cut, seeded, peeled, cut in chunks and boiled until tender for about 10 to 20 minutes. Simply dot with butter and add a little salt, pepper and a squeeze of orange or lemon juice if desired or glaze the cooked pieces in a frying pan with a little butter, brown sugar and seasoning. Winter squash is delicious in soup and stew, steamed, puréed, batter fried or braised.

- Spaghetti squash is best baked in a moderately hot oven for 1 to 1 1/2 hours, depending on its size. Pierce it with a knife in 2 or 3 places first. When the skin can be easily depressed, remove from heat, halve lengthways, discard seeds and carefully remove the flesh, fluffing it up with 2 forks. The slightly crunchy, bland, spaghetti-like flesh is excellent served with a variety of pasta toppings, such as cheese and butter, tomato sauce and clam sauce, or combined with mornay sauce and baked au gratin. Try leftovers cold with a vinaigrette.

➠ Suggestions

- Young, tender summer squash, especially zucchini, chayote and cocozelle, can be eaten raw in salad or with dips.

• Try summer squash lightly steamed or stir-fried with a hint of butter, lemon and garlic, fried in batter, prepared in a variety of Mediterranean recipes from ratatouille to baked Parmesan, in soups, quiche, curries, stuffed and baked, or even pickled or marinated.

• Summer squash is also tasty grilled in the oven or on a barbecue with a little olive oil and herbs; slice lengthwise if zucchini-shaped or crosswise if rounded or leave whole if very small. Cut in chunks for kebabs.

• Use a basic pancake or fritter recipe and add 250 mL (1 cup) of fresh grated summer squash, a small minced onion and your favourite fresh herb. Form small patties and fry them in a small amount of oil until golden brown and serve them with yogourt or sour cream. Alternatively, add 250 mL (1 cup) cooked, puréed winter squash to the basic mixture and add nutmeg, cinnamon and ginger to taste. Serve with honey or maple syrup.

• Bake small winter squash such as acorn, butternut, gem or sweet dumpling until tender (30 to 60 minutes); first halve or quarter (unless they are very small, in which case just prick the skin with a knife a few times) and remove the seeds, then brush on a little butter or oil, salt, pepper and a little nutmeg. A touch of maple syrup or brown sugar adds a great flavour.

• For a fancier dish, after about 20 minutes of baking, try hollowing out some of the flesh of each squash half, dicing it and adding it to a stuffing, such as seasoned rice, mushrooms, raisins and pine nuts or simply top with a little grated cheese and breadcrumbs, then continue baking until very tender. Apple sauce is another good filling, especially when accompanying lamb or pork chops or roast chicken.

• Mash cooked winter squash with a little butter, a spoonful of honey, finely chopped orange zest, a little finely grated fresh ginger, a pinch of nutmeg and salt and pepper to taste.

• To make winter squash Indian-style: Sauté a minced onion, 2 minced garlic cloves, a little grated ginger, turmeric, and curry powder in a small amount of butter or ghee. After a minute or two, add a squeeze of lemon juice, a pinch of salt, and some peeled, chopped squash. Stir, then add just a little water. Cover and continue cooking until the squash is tender, adding a bit more water if it seems too dry.

• Cooked squash is good in a variety of desserts, including sweet breads, muffins, pies and puddings.

• For a delicious winter squash soup: In a large pot, sauté 2 large chopped onions and 2 minced garlic cloves in a little butter until tender but not browned. Add 2 small or 1 medium orange-fleshed winter squash, peeled, seeded and chopped in chunks, along with some grated fresh ginger, 750 mL (3 cups) water and 2 or 3 chicken stock cubes. Boil, then simmer until tender, about 20 to 30 minutes. Purée, then add 250 mL (1 cup) apple or fresh orange juice and season with salt and pepper to taste. Reheat before serving, adding a little more water if necessary. Garnish with fresh herbs or croutons if desired.

Winter squash soup ↓

Nutritional Value

Butternut squash

(per 100 g = 3 1/2 oz = 1/2 cup)

			RDA**	%
calories*	(kcal)	69.5		
protein*	(g)	1.9		
carbohydrates*	(g)	17.9		
fat*	(g)	0.1		
fibre	(g)	1.9	30	6.3
vitamin A	(IU)	6560	3300	198.8
vitamin C	(mg)	8	60	13.3
thiamin	(mg)	0.05	0.8	6.3
riboflavin	(mg)	0.13	1.0	13.0
niacin	(mg)	0.7	14.4	4.86
sodium	(mg)	1	2500	0.04
potassium	(mg)	624	5000	12.5
calcium	(mg)	41	900	4.6
phosphorus	(mg)	74	900	8.2
magnesium	(mg)		250	
iron	(mg)	1.05	14	7.5
water	(g)	81.6		

*RDA variable according to individual needs
**RDA recommended dietary allowance (average daily amount for a normal adult)

Fried zucchini

(serves 4)

from the Montreal restaurant Trattoria Trestevere

Ingredients		
6-8	zucchini	
125 mL	milk	1/2 cup
250 mL	flour	1 cup
	vegetable oil for frying	
	salt and pepper to taste	
15 mL	parsley, chopped finely	1 TBSP
	lemon wedges	

Instructions

- Slice the zucchini like thin french fries and sprinkle on a bit of salt. Leave the slices spread out on a paper towel for about an hour to dry out.
- Put the zucchini in a shallow bowl and sprinkle on the milk. Let sit about 2 minutes, stirring a few times. Squeeze out excess milk.
- Heat the oil to about 150 °C (300 °F).
- Toss the zucchini pieces in the flour, then put them in a large strainer and shake off the excess flour.
- Put the zucchini pieces in the hot oil, preferably in a frying basket, so they are easily removed. As the slices turn golden brown, they will rise to the surface. It should take about 2 minutes to cook them all.
- Drain off excess oil on paper towels. Sprinkle on salt, pepper and parsley and serve with lemon wedges.

Chayotes au gratin

(serves 2-4)

Ingredients		
2	chayotes	
60 mL	milk	1/4 cup
15 mL	butter or margarine	1 TBSP
30 mL	grated cheese	2 TBSP
250 mL	breadcrumbs	1 cup
	salt and pepper to taste	
2-4	strips of bacon (optional)	
1	small onion, minced	
1	garlic clove, minced	
15 mL	parsley, minced	1 TBSP
30 mL	grated Parmesan cheese	2 TBSP

Instructions

- Place whole chayotes in a pot and add water to cover. Bring to a boil, then simmer about 20 minutes.
- Drain, cut each chayote in half lengthwise and discard the central seed. Scoop out the flesh, leaving the skins intact with a little flesh on them to help keep their shape.
- Mash the chayote flesh, discard any excess water, and mix it well with the milk, butter, grated cheese, 3/4 of the breadcrumbs and salt and pepper to taste.
- Chop the bacon into small pieces and sauté it with the onions, garlic and parsley. Mix this with the other ingredients, and stuff the chayote skins with equal amounts.
- Mix the grated Parmesan cheese with the remainder of the breadcrumbs and sprinkle evenly over the stuffed chayotes.
- Bake at 175 °C (350 °F) about 20 minutes or until nicely browned.

← **Chayotes au gratin**

Seville and blood oranges

↑ Blood orange

BLOOD ORANGES ARE VERY LOW IN CALORIES
EXCELLENT SOURCE OF VITAMIN C
VERY GOOD SOURCE OF VITAMIN A

• Very popular in Europe, the "sour" Seville orange and the "sweet" blood orange are beginning to gain recognition in North America at long last. The blood orange is an attractive and tasty hybrid that originated in Europe in the mid-19th century. These two unusual oranges are grown mainly in southern Europe (especially Spain and Italy) and North Africa, though some are now being cultivated in the citrus-growing areas of the U.S.

Origin

• The orange, from the important genus citrus of the Rutaceæ family, is closely related to the pomelo, grapefruit, lemon, lime, mandarin and the various tangors. All citrus fruits, known for their ability to cross breed easily, are presumed to have evolved naturally over millions of years from one or more types of wild citrus, indigenous to Southeastern Asia.

• Oranges are usually classified as sour or sweet and some botanists believe that sour oranges are the ancestors of sweet ones. The Moors brought sour oranges from India back to Persia in the 10th century, then later introduced them to Northern Africa and Spain. The Chinese were among the first to cultivate and write about sweet oranges. Sweet oranges, not known in Europe before the 15th century, were improved and planted widely by the Portuguese.

Varieties and description

• The Seville orange is a cooking orange. On the small side, it has a yellow or green tinge to its thick, orange,

Seville orange ↑

dimpled rind, which sometimes shows a small amount of russeting. Its pale orange flesh, filled with seeds, is not very juicy and is considered too bitter to be eaten fresh.

• The two main types of blood oranges are the **full bloods**, with their skin and flesh splashed with deep blood red patches, and the **semi-bloods,** with orange skin and red-speckled flesh. This tempting orange is deliciously sweet, yet has a full-bodied citrus flavour. Most types are seedless.

Availability

• American Seville oranges are harvested in January and February, while those from Spain are available in February and March. Availability is usually limited, as the crop is small.

• European and African blood oranges are in season from November to March. A small crop of American ones is available during the winter months.

Selection

• All citrus should feel heavy for its size and fairly firm. Citrus is picked when fully ripe and can be used immediately.

Storage

• Seville oranges can be refrigerated in the crisper for about 2 weeks. Blood oranges also need refrigeration. They are best eaten within a week or 10 days.

Preparation

• The peel and flesh of Seville oranges is traditionally used to make the finest and most delicious marmalade. There is no better orange for this purpose, as the Seville orange contains a lot of pectin and the right level of acid.

• Blood oranges are good eating oranges. They produce a beautifully coloured juice as well. They are excellent as a garnish, peeled and thinly sliced crosswise. Arrange decoratively in a circle around the outer edge of a fruit salad or a salad platter, to give a flamboyant colour.

Suggestions

• Seville oranges are excellent in duck à l'orange and in sauces used for other types of poultry and for fish fillets. The flavour is richer than other oranges' and perfectly enhances sweet and sour or barbecued pork and spicy beef with orange peel.

• Use the juice of a Seville orange in a vinaigrette instead of lemon or vinegar.

• A little Seville orange juice added to club soda makes a refreshing drink.

• Most commercially produced orange-flavoured liqueurs are made with Seville oranges. To make a delicious orange liqueur at home, mix together 1 bottle of white wine, the peel of one Seville orange and juice of two, 90 mL (6 TBSP) sugar and 15 mL (1 TBSP) dried chicory root or whole coffee beans. Let the mixture stand in a decanter or a large bottle in a cool, dark place for 2 weeks before using. Strain, then keep refrigerated.

• Peel and roast whole blood oranges with a chicken for about 30 minutes, turning often to brown and glaze them in the poultry juices.

• For a simple and elegant dish, dust fillets of sole or trout lightly in flour and sauté in a little butter for a few minutes, adding a dash of salt and pepper. Set aside on a heated platter, and sauté the segments of 2 blood oranges (after removing the membrane) with 1 finely minced garlic clove, 2 minced scallions or dry shallots, a squeeze of lemon, seasoning and a little splash of white wine. Arrange on the fish platter, garnishing with parsley, and serve with rice cooked with a little finely grated orange zest.

• For a tasty dessert, segment 6 blood oranges and place in an oven proof dish with a sprinkling of cinnamon. In a bowl, mix 80 mL (1/3 cup) each brown sugar, chopped walnuts and flour. Cut in 30-45 mL (2-3 TBSP) margarine, then work mixture with your fingers until well crumbled. Spread over oranges and bake at 200 °C (400 °F) 30 minutes or until browned. Sprinkle on bittersweet chocolate chips and bake 1 minute.

Nutritional Value

Blood orange

(per 100 g = 3½ oz)

			RDA**	%
calories*	(kcal)	40		
protein*	(g)	0.8		
carbohydrates*	(g)	9.9		
fat*	(g)	0.2		
fibre	(g)	0.4	30	1.3
vitamin A	(IU)	500	3300	15.2
vitamin C	(mg)	43	60	72
thiamin	(mg)	0.07	0.8	8.8
riboflavin	(mg)	0.04	1.0	4
niacin	(mg)	0.4	14.4	3
sodium	(mg)	2	2500	0.08
potassium	(mg)	162	5000	3.2
calcium	(mg)	21	900	2
phosphorus	(mg)	20	900	2
magnesium	(mg)		250	
iron	(mg)	0.3	14	2.2
water	(g)	88		

*RDA variable according to individual needs
**RDA recommended dietary allowance (average daily amount for a normal adult)

Mandarin hybrids

↑ Moroccan clementine

EXCELLENT SOURCE OF VITAMIN C
GOOD SOURCE OF B VITAMINS
VERY LOW IN CALORIES

Origin

• It is extremely difficult to categorize mandarins, which have a similar origin, though a more recent history, than the orange. Though closely related to oranges, mandarins have a thinner, looser skin that peels off more easily. Cultivated in China and Japan for several thousand years, these attractive orange fruit are named for Chinese mandarins, the government officials who are traditionally dressed in bright orange clothing.

• Tangerines are actually not a variety, but are named after the Moroccan city Tangier, which was once the main shipping point for the tasty little citrus fruit that came to be associated with Christmas treats. The name has become so widely used that it has generally been adopted by both growers and the public to describe the hybrids created by crossing mandarins with different types of oranges. Tangerines were first introduced to Europe at the beginning of the 19th century and to the United States about 50 years later. The main citrus producing areas are Florida, California, Arizona and Texas.

Varieties and description

• Most varieties, round in shape and slightly flattened at the top and bottom, are much sweeter than oranges. They are graded from small to super colossal.

• The **clementine,** a superb hybrid created by Father Clément Rodier in Algeria at the beginning of the century, is a sweet, deep reddish-orange fruit that comes in a variety of sizes. It has a loose, at times puffy, pebbled rind that peels very easily. It is usually seedless. Clementines are sometimes called Algerian tangerines.

• The **Fairchild tangerine,** a smooth-skinned, deep orange, medium to large fruit, usually contains quite a few seeds. It has a very rich, sweet flavour.

• **Honey tangerines,** also called Murcotts after Charles Murcott Smith who developed them, are more firm than most tangerines. Exceptionally flavourful and juicy, they often contain quite a few seeds. Both skin and flesh are deep orange.

• **Tangelos,** sweet yet tangy, result from crossing a tangerine with a grapefruit. The **Orlando tangelo** is a large to extra-large fruit that has a tight-fitting, lightly pebbled, orange skin and contains some seeds. The large **Minneola tangelo,** with its knob-like protrusion at the stem end, is a beautiful deep red-orange fruit that is easily peeled. It has very few seeds.

Spanish clementine ↓

• The **Kinnow mandarin,** a medium to jumbo fruit, has a thin, yellowish orange, smooth skin. It's fruity, sweet flesh contains many seeds.

• Among the many mandarin hybrids grown commercially in the U.S. are the **Satsuma, Dancy tangerine, Temple mandarin, Seminole tangelo, Kara mandarin, Fremont tangerine, Fortune tangerine** and **Wilking mandarin.**

Availability

• The American clementine is available from November to February, with peak supplies at Christmas time. This is usually when most of the delicious tiny Moroccan and medium-sized Spanish clementines are imported.

• Fairchild tangerines are available in November and December.

• American honey tangerines are available in January and February, while those from South America, particularly Brazil, are in season from July to October.

• Orlando tangelos are found from December to March, while Minneolas can be bought from January to April.

• Kinnow mandarins are in season from January to May.

Selection

• As with other citrus fruits, a heavy fruit indicates juiciness. Puffy skin is quite normal, but the skin should not be completely detached from the fruit. Green colour in the skin does not indicate that the fruit is unripe or that the taste has been affected, but that the weather was too

↓ Fairchild tangerine

Honey tangerine ↑

↓ Orlando tangelo

↑ Kinnow mandarin

warm when the fruit was developing. Citrus fruit requires warm days and cool evenings to produce orange-coloured skins; citrus fruit grows better in sub-tropical climates than in the tropics.

Storage

• More perishable than oranges, mandarins, tangelos and tangerines should be kept at room temperature for only a day or two, or refrigerated in a plastic bag for about a week.

Preparation

• Wash all fruit just before using. Tangerine zest, milder than orange zest, is useful in many recipes, and can be dried in a slow oven or frozen for later use. (Add a little citrus zest to tea or coffee for a pleasant, spicy flavour.)

• These easy-to-eat fruit are usually enjoyed fresh, but add colour and flavour to cooked dishes and desserts. If whole tangerine segments are required in a recipe, slit open the narrow side of each one and pop out any seeds.

Suggestions

• A large bowl of clementines, served for dessert or a snack, will disappear in minutes, especially when children are around. Include a couple in lunch boxes.

• Tangerine juice, especially when made from honey tangerines (Murcotts), is refreshing and delicious.

• Add tangerine segments to chicken salad, shrimp salad, cole slaw, cottage cheese salad, fruit salad or to a regular vegetable salad.

• Tangerine segments are very tasty used in sweet and sour chicken or pork dishes.

• Add tangerine segments to puddings, gelatins, sweet soufflés, rice pudding or other desserts for added flavour and colour. They are excellent as a topping for French open-faced tarts.

• To perk up a carrot or squash soup, add the juice of a few tangerines and a little finely chopped zest.

• Make tangerine sorbet and fill tangerine shells with it after it has been partially frozen.

• For an appetizing dessert, bake a bottom pie crust until golden brown. In a bowl, blend until smooth 1 package cream cheese with 2 small containers plain yogourt, 60 to 80 mL (1/4 to 1/3 cup) sugar, 5 mL (1 tsp) vanilla and a sprinkle of cinnamon. Peel and segment 6 tangerines, removing any pips. Stir tangerines and 125 mL (1/2 cup) raisins or small seedless grapes into mixture, fill pie shell and refrigerate for a few hours before serving.

Nutritional Value

Clementine

(per 100 g = 3½ oz)

			RDA**	%
calories*	(kcal)	34		
protein*	(g)	0.5		
carbohydrates*	(g)	8.7		
fat*	(g)	0.1		
fibre	(g)	0.3	30	1.0
vitamin A	(IU)	83.3	3300	2.5
vitamin C	(mg)	32	60	53.3
thiamin	(mg)	0.05	0.8	6.3
riboflavin	(mg)	0.03	1.0	3.0
niacin	(mg)	0.3	14.4	2.1
sodium	(mg)	3.0	2500	0.1
potassium	(mg)	139	5000	2.8
calcium	(mg)	20	900	2.2
phosphorus	(mg)	16	900	1.7
magnesium	(mg)		250	
iron	(mg)	0.2	14	1.4
water	(g)	90.4		

*RDA variable according to individual needs
**RDA recommended dietary allowance (average daily amount for a normal adult)

Tangerine and spinach salad

(serves 4)

Ingredients		
1	red onion	
4	tangerines	
1	package or bunch very fresh spinach	
1	small handful almonds, slivered and toasted	
	Dressing	
45 mL	fresh lemon juice	3 TBSP
30 mL	walnut oil	2 TBSP
5 mL	aniseed or fennel seeds	1 tsp
	salt and pepper to taste	

Instructions

- Slice red onion (or sweet onion) in very thin rings.
- Mix the dressing and marinate the onions in it for about an hour.
- Peel tangerines and remove all white pith and any seeds. Cut each segment in 2 or 3 pieces if the fruit is large, or leave whole.
- Wash and dry spinach thoroughly, discarding stems. Tear into bite-sized pieces.
- Mix onions and dressing with spinach and tangerines just before serving, and garnish with toasted almond slivers.

Kumquat

LOW IN CALORIES
VERY GOOD SOURCE OF VITAMIN C
CONTAINS PHOSPHORUS AND IRON

Origin

• Known to China and Japan for several thousand years, the exquisite little kumquat was introduced to Europe only in the mid-19th century by the British horticulturalist, Robert Fortune. Though a member of the Rutaceæ family, the kumquat was removed from the citrus genus earlier this century and a new genus, Fortunella, was created, to emphasis its botanical differences with other types of citrus.

• The kumquat, meaning "golden orange" in Chinese, was introduced to Florida and later to California soon after it was brought to Europe. The American crops have been used mainly for processing in the past, but more of the fresh fruit is now being brought to market. Kumquats are also grown in Brazil, Peru, southern Europe, Israel and parts of Africa, as well as in the Orient.

Varieties and description

• The oval-shaped nagami kumquat has a sweet rind and tart flesh; the round Marumi kumquat is quite sharp; while the Meiwa kumquat, a hybrid of these two, is quite sweet.

• The bright golden yellow to orange kumquat is entirely edible, though most people prefer to discard the pale green or white seeds. It has a thin, sweet rind and fragrant, at times acidic, pale orange flesh. Usually 2 to 5 centimetres (1 to 2 inches) long, kumquat varieties differ in shape and flavour.

• Among the many kumquat hybrids are the limequat, citrangequat, lemonquat, orangequat and the Filipino calamondin, or scarlet lime.

Availability

• Kumquats are available year round, as there are so many producer countries. In North America it is most often seen from November to April, the season for the Florida and California crops.

Selection

• Look for firm, bright-skinned, unblemished kumquats. Those sold with their dark green leaves and twigs still attached are great for fruit bowls and table decorations.

Storage

• Kumquats will keep at room temperature for 5 or 6 days. They can be stored for longer periods — up to 3 or 4 weeks — in the refrigerator, loosely wrapped.

Preparation

• Wash kumquats well. They can be blanched and washed in cold water to soften their skins, if desired. Slice thinly, halve or quarter and remove pips or leave whole, according to the recipe.

• Kumquats are often used for marmalade, brandied preserves, and pickles or even poached in syrup, though their bittersweet flavour is delightful when fresh or lightly cooked in a sauce.

Suggestions

• Kumquats provide a perfect tangy enhancement to duck, chicken, turkey, lamb and pork dishes.

• Use thinly sliced kumquats in fancy fruit punch or to garnish cocktails.

• Excellent in fruit salads, kumquats are also decorative and tasty in vegetable, chicken or seafood salads.

• Kumquats add a great flavour to sweet and sour sauces.

• Use finely sliced and halved kumquats in fruit cakes or muffins. They can be seeded and puréed for cake icing, sorbet or other desserts.

Kumquat tart

Ingredients		
1	tart crust	
	Filling	
125 mL	sugar	1/2 cup
	juice of 3 oranges	
	squeeze of lemon juice	
	zest of one orange	
45 mL	butter	3 TBSP
3	eggs	
	Topping	
20-24	kumquats	
60 mL	sugar	1/4 cup
45 mL	Grand Marnier (optional)	3 TBSP

Instructions

- Use a flat French tart crust, and bake it until it is golden brown.
- Make the filling in a double boiler: Over moderate heat add the sugar, orange and lemon juice, zest and butter. Whisk continuously until the sugar is well dissolved.
- Beat the egg and add it very slowly to the mixture. Continue whisking until the orange cream has thickened, but don't let it boil.
- Remove from heat, let cool, then spread evenly on the tart.
- Slice the kumquats very thinly, removing any seeds. Put them in a large frying pan, cover with water and add the sugar and the Grand Marnier (or another orange-flavoured liqueur). Heat until sugar is dissolved, then simmer about 5 minutes, until the fruit is tender.
- Drain the kumquats very well, then arrange decoratively as a topping for the tart.
- Reserve the syrup for fruit salad, if desired.

Nutritional Value

Kumquat

(per 100 g = 3 1/2 oz)

			RDA**	%
calories*	(kcal)	48		
protein*	(g)	0.4		
carbohydrates*	(g)	11.3		
fat*	(g)	0.5		
fibre	(g)	1.5	30	5.0
vitamin A	(IU)	42	3300	1.2
vitamin C	(mg)	40	60	66.7
thiamin	(mg)	0.09	0.8	11.3
riboflavin	(mg)	0.06	1.0	6.0
niacin	(mg)	0.5	14.4	3.5
sodium	(mg)	111	2500	4.4
potassium	(mg)	156	5000	3.1
calcium	(mg)	16	900	1.8
phosphorus	(mg)	65	900	7.2
magnesium	(mg)		250	
iron	(mg)	0.8	14	5.7
water	(g)	86.7		

*RDA variable according to individual needs
**RDA recommended dietary allowance (average daily amount for a normal adult)

Pomelo and uglifruit

↑ Pomelo

VERY RICH IN VITAMIN C
CONTAINS SOME B VITAMINS
LOW IN CALORIES AND SODIUM

Origin

• The pomelo or pummelo (sometimes called shaddock), a citrus fruit native to Southeast Asia, has been a favourite fruit for thousands of years in China, where it is believed to be an aid to digestion.

• The pomelo is the ancestor of the grapefruit. A British trader called Captain Shaddock brought pomelo seeds to the Barbados in the last decade of the 17th century and from there they were spread by man or birds to other islands, notably Jamaica. It was in Jamaica that the grapefruit, which is probably a naturally occurring mutation of the pomelo, was first discovered in the early 19th century.

• Pomelos are grown in tropical and subtropical regions throughout the Orient and Africa, as well as in the Caribbean, South America, and to a lesser extent, in the citrus-growing areas of the United States.

• The uglifruit (pronounced oo-glee) is a hybrid citrus that may have occurred naturally, as most types of citrus crossbreed extremely easily. It is believed to be a cross between a pomelo or grapefruit and a tangerine. It is native to Jamaica and most uglis are still grown there, though Brazil is beginning to produce them as well. Canadians consume most of the world's crop.

Varieties and description

• There are a number of varieties of pomelos. Globular or pear-shaped, from the size of a grapefruit to the size of a basketball, all pomelos have a very thick, aromatic rind. Their skin is usually yellow, tinged with green or pink and can be smooth or bumpy. Much sweeter than grapefruit, their juicy, white, pale yellow or pink flesh is generally more coarsely textured. Most varieties have seeds.

• Uglifruit has a rough green skin, mottled with yellow or orange and sometimes russeted in spots — at first sight not very appealing. Ranging from the size of a large orange to a large grapefruit, the uglifruit has a loose skin that peels easily, like a tangerine's. Its pink-to-orange, juicy, fragrant flesh, which contains few seeds, combines the best flavour elements of a sweet tangerine and a tangy grapefruit.

Availability

• Pomelos are available from various producer countries from August to January.

• Uglifruit are in season from December to May.

Selection

• Choose firm, not hard pomelos and uglifruit that feel heavy for their size. A large size is not necessarily an indication of a better fruit. It is difficult to say how to choose a good pomelo, as they can vary enormously from rather dry and bland to juicy and extremely tasty. Pay no attention to scars or marks on uglifruit, but avoid those that look dried out at the stem end.

Storage

• Both pomelos and uglifruit can be left at room temperature if they are to be eaten within a day or two. They can be refrigerated for about a week, but are juiciest when eaten as fresh as possible.

Preparation

• To eat a pomelo, use a sharp knife to peel off all of the outer skin and the thick inner pith. Halve the fruit, then separate into segments, removing the rather tough membrane surrounding each segment and the seeds, if any. If the flesh seems a little dry, serve the fruit in a light syrup or steeped in a little fruit juice. The thick rind (including the skin and white pith) can be reserved, if desired, to make a preserve or candied peel — a traditional Chinese treat.

• Uglifruit are best peeled like a tangerine and enjoyed fresh.

Suggestions

• For a delicious breakfast or a light dessert, peel a pomelo, removing the membrane and seeds and place the segments flat in a baking dish. Sprinkle on a little brown sugar and place under the broiler for a few minutes.

• Pomelos and uglifruit are most often enjoyed fresh, alone or in a variety of salads, including fruit, vegetable, cottage cheese and seafood salads.

• Try heating pomelo segments for a minute or two with a splash of Pernod and honey and use as a filling for crêpes or a topping for ice cream.

• For a gourmet salad, marinate minced scallions or shallots and small pieces of pomelo or uglifruit in a zesty vinaigrette for an hour, then toss in a salad composed of lettuce, endive and radicchio or thinly sliced red cabbage. Garnish with slivers of avocado and finely chopped parsley.

• Another excellent salad can be made by combining 1 or 2 pomelos or uglifruits in seeded sections with 1 bunch lightly cooked, bite-sized pieces of broccoli, 1 cooked and julienned chicken breast, and 1 diced avocado. Toss in a garlic vinaigrette and arrange on a bed of lettuce.

Uglifruit ↑

Nutritional Value

Pomelo

(per 100 g = 3½ oz)

			RDA**	%
calories*	(kcal)	39		
protein*	(g)	0.7		
carbohydrates*	(g)	9.5		
fat*	(g)	0.3		
fibre	(g)	0.4	30	1.3
vitamin A	(IU)	50	3300	1.5
vitamin C	(mg)	53	60	88.3
thiamin	(mg)	0.05	0.8	6.3
riboflavin	(mg)	0.02	1.0	2.0
niacin	(mg)	0.3	14.4	2.1
sodium	(mg)	1	2500	0.04
potassium	(mg)	235	5000	4.7
calcium	(mg)	27	900	3.0
phosphorus	(mg)	22	900	2.4
magnesium	(mg)		250	
iron	(mg)	0.5	14	3.6
water	(g)	88.9		

*RDA variable according to individual needs
**RDA recommended dietary allowance (average daily amount for a normal adult)

Nutritional Value

Uglifruit

(per 100 g = 3½ oz)

			RDA**	%
calories*	(kcal)	44		
protein*	(g)	0.7		
carbohydrates*	(g)	11.2		
fat*	(g)	0.1		
fibre	(g)	2	30	6.7
vitamin A	(IU)	116.7	3300	3.5
vitamin C	(mg)	43	60	71.7
thiamin	(mg)	0.07	0.8	8.8
riboflavin	(mg)	0.03	1.0	3
niacin	(mg)	0.3	14.4	2.1
sodium	(mg)		2500	
potassium	(mg)		5000	
calcium	(mg)	42	900	4.7
phosphorus	(mg)	20	900	2.2
magnesium	(mg)		250	
iron	(mg)	0.4	14	2.9
water	(g)	87.5		

*RDA variable according to individual needs
**RDA recommended dietary allowance (average daily amount for a normal adult)

Cranberry

VERY LOW IN CALORIES AND SODIUM
CONTAINS VITAMIN C AND IRON

Origin

• The cranberry, native to North America, is a member of the heath family, Ericaceæ, which also includes the blueberry, huckleberry, bilberry and lingonberry. The natural habitat of the cranberry is a swamp or bog in a cool climate. The various Indian names meant "sour berry", but the settlers called it crane berry, possibly because it was a preferred food of cranes or because the plant in flower looks like one of those elegant birds.

• North American Indians used cranberries extensively for dyes, medicine and food — especially pemmican, a dried meat and cranberry preparation. Cranberries were taken to Europe by the mid-16th century by returning explorers, but did not become immediately popular among Europeans or American settlers. American sailors and loggers in the 19th century used cranberries to make a drink, that helped prevent scurvy, though they probably didn't know that its medicinal value was due to its high content of vitamin C.

• Cranberries were gathered from the wild until about 150 years ago. Today they are a very important crop in Massachusetts and are cultivated as well in New Jersey, Wisconsin, Oregon, Washington, British Columbia and Quebec.

Description

• These festive, firm, light-to-deep-red berries have a smooth, glossy skin. Oval in shape and generally 1.3 centimetres (1/2 inch) in diametre and 1.3 to 2.5 centimetres (1/2 to 1 inch), the berries contain tiny, soft, edible seeds. Cranberries have a tart flavour that compliments a great variety of foods.

Availability

• Fresh cranberries are available from September to January. They usually are packaged in 350-gram (12-ounce) cellophane bags, which yields about 750 mL (3 cups).

Selection

• Cranberries have a very high acidity content, which gives them a long shelf life. Packaged cranberries are always sold in good condition, unless there has been a problem with their refrigeration.

Storage

• Keep fresh cranberries refrigerated for up to 4 weeks. Freeze unwashed berries, well sealed in a double plastic bag, and they will last about a year. Use frozen berries directly in a recipe without thawing them first.

• Leftover cranberry sauce will keep well for a few days in the refrigerator, but is best frozen for later use in conveniently sized air-tight containers.

Preparation

• Wash cranberries well just before using, discarding any soft, withered or bruised berries and any stems. Cranberries need to be sweetened with sugar, honey or maple syrup. Their high pectin content makes them an ideal fruit for jelly or fruit chutney.

Suggestions

• Besides the popular cooked cranberry sauce and jelly used to accompany roasted poultry, meat and fish, cranberries can be used in an equally delicious uncooked relish. Grind or mince 1 package of berries with the zest and pulp of an orange and mix it with 190 to 250 mL (3/4 to 1 cup) of sugar. Add cinnamon if desired. Let the mixture sit overnight to blend the flavours. It will keep for up to 2 weeks in the refrigerator.

- Cranberries are excellent mixed with apples or pears in pies, tarts, compotes and crumbles. Baked apples and pears are very appetizing stuffed with chopped cranberries and maple syrup.

- Try cranberry fruit cake, cranberry nut loaf, cranberry upside-down cake or cranberry muffins. Cranberry sorbet is great for slimmers, while cranberry cheese cake makes a beautiful party dessert.

- For a tasty cranberry chicken dish, cut a chicken in quarters and brown on both sides in a deep frying pan. Add 1 minced onion, grated or slivered ginger to taste and the zest of 1 orange and the juice of 2. Simmer, covered, for about 20 minutes, turning the pieces of chicken after 10 minutes. Add 250 mL (1 cup) fresh cranberries and 125 mL (1/2 cup) of sugar and simmer, uncovered, for about 10 minutes.

- For a decorative look and a zesty flavour, add cooked and sweetened cranberries to fruit, chicken or vegetable salads or use them to garnish a variety of puréed vegetable side dishes, such as potato, yam, squash or turnip.

- Stuff halved and cored pears with cranberry sauce, sprinkle on orange juice and cinnamon and bake 20 minutes in medium oven.

Nutritional Value

Cranberry

(per 100 g = 3½ oz)

			RDA**	%
calories*	(kcal)	46		
protein*	(g)	0.4		
carbohydrates*	(g)	10.8		
fat*	(g)	0.7		
fibre	(g)	1.4	30	4.7
vitamin A	(IU)	41.6	3300	1.2
vitamin C	(mg)	11.0	60	18.3
thiamin	(mg)	0.03	0.8	3.8
riboflavin	(mg)	0.02	1.0	2.0
niacin	(mg)	0.1	14.4	0.7
sodium	(mg)	2.0	2500	0.08
potassium	(mg)	82	5000	1.6
calcium	(mg)	14	900	1.6
phosphorus	(mg)	10	900	1.1
magnesium	(mg)		250	
iron	(mg)	0.5	14	3.6
water	(g)	87.9		

*RDA variable according to individual needs
**RDA recommended dietary allowance (average daily amount for a normal adult)

Cranberry sauce in squash boats

(serves 4-8)

Ingredients

1	package fresh cranberries	
1	orange	
1	apple	
	zest of one orange	
250 mL	sugar	1 cup
75 mL	water	1/4 cup
2-3	small squash	

Instructions

- Use fresh or frozen cranberries; wash them well, discarding any stems or soft berries.

- Peel and segment the orange, removing the membrane. Peel and core the apple and dice both fruits.

- Combine all the ingredients in a saucepan, bring to the boil, then simmer for 6 to 10 minutes, until the cranberry skins begin to pop.

- Cool, then add a little more sugar if necessary.

- Halve the squash crosswise and discard the seeds. Bake until tender in a greased baking pan for 40 to 50 minutes, starting upside down and turning upright after about 25 minutes. Fill the hollows with cranberry sauce.

- Serve with turkey, chicken, lamb or pork chops.

Pears

↑ Seckel

EXCELLENT SOURCE OF FIBRE
SOURCE OF VITAMIN C AND RIBOFLAVIN
VERY LOW IN CALORIES AND SODIUM

Origin

• Pears are members of the large Rosaceæ family. Originally from Northeastern or Central Asia, pears have been known to man since prehistoric times and cultivated for nearly 4,000 years, though it is probable that the earliest cultivated types were Asian apple pears. The pear was known to ancient civilizations from the Chinese to the Romans, who helped spread pear trees throughout Europe.

• Until the 16th century, pears were always cooked, but when new varieties were hybridized and found to be delicious uncooked, pears gained enormously in popularity. In the 17th and 18th centuries, the gardeners of many European noblemen competed to crossbreed the perfect pear, which they thought should have a buttery soft texture. Most of today's favourite pears are derived from those improved varieties. They are now grown throughout Europe, with France as the leading producer.

• Pears, which today are cultivated in great quantities in western regions of North America, were first taken there by French and Spanish missionaries as well as by the colonial settlers. They were later introduced to South Africa, Australia, New Zealand and the temperate zones of South America, where they also grow well.

Varieties and description

• Pears cross breed easily and sometimes mutate, resulting in several thousand varieties. All varieties have white or creamy-coloured flesh, edible skin and most are aromatic.

• **Bartlett** and **red Bartlett** pears, known as **Williams** in Europe, are bell-shaped. Bartletts turn from green to yellow with a pink blush, while red Bartletts (a hardier variety developed in the United States) have an attractive bright crimson skin or red streaks on a green background. Both varieties taste alike.

• **Seckels** are small, bite-sized, excellent tasting, short-necked pears, usually dark green with a dull red blush or dark red. A mutation that occurred naturally in Pennsylvania, they were planted extensively by a man named Seckel.

• **Bosc** pears, similar in appearance to European **Conference** pears though much juicier and less granular in texture, are long-necked, with golden brown russeted skin.

• **Anjou** pears are almost oval, with very little indentation at the neck. They are light green or yellowish green and extremely juicy and buttery soft, with a hint of spiciness.

• **Comice** pears, usually large, round and very short-necked, have greenish yellow skins, at times blushed with red or russeted. These superb pears are very sweet and fragrant. Originally a French variety, they have been grown in North America for more than 100 years.

• **Forelles** are rather small and have an elongated bell-shape. They are golden yellow, with a bright red blush and distinctive red freckles.

Anjou ↓

Red Bartlett →

↑ Bosc

• **Clapp** pears are large, oval and greenish yellow, with a red blush on the side most exposed to the sun.

• **Rocha** pears, a Portuguese variety, are lightly russeted over a golden yellow background, with pronounced russeting at the stem end. They are medium-sized and round, with a short neck. The Rocha turns from firm and crisp to soft and mouth-wateringly juicy.

• **Passe-Crassane** is a variety developed by a Normandy grower more than 100 years ago and today is one of the most popular winter pears in Europe. Cultivated extensively in France, it is large, plump, round and golden coloured, with a rougher skin than summer pears. It is extremely juicy and flavourful, with a slightly more granular texture than most American pears.

• **Packham** pears, developed in Australia, are similar to Bartletts in colour and flavour, though not as regular in shape.

Availability

• The Clapp pear is an early summer variety, found from July to September. Bartletts, red Bartletts and Seckels are available from July to December. Bosc pears have a long season, from August to May, as do Anjous, from October to July. Comice pears can be found from August to March and Forelles from October to February. Portuguese Rocha pears are in season from the end of September to January. French Passe-Crassane pears are available from November to the beginning of February. Packham pears can be found from November to August, from various producer countries.

Selection

• Pears are one of the few fruits, along with bananas and avocados, that do not mature well if left to ripen on the tree. The fruit forms "stone" or "grit" cells when it ripens on the tree, leading to a sandy or granular texture. However, if the pear is picked green, it will ripen without forming these cells, resulting in a smooth, soft and pleasant texture.

• Most stores sell firm to hard pears because they bruise less easily than ripe ones, making it difficult to find fruit that can be enjoyed the same day. Fancy varieties come well packed to protect the fruit.

↓ Comice

Forelle ↓

Storage

- Pears are best left at room temperature until they yield to gentle pressure at the stem end. Most varieties do not change colour when they ripen (i.e. some varieties are green when ripe), so a slight softening is the best indication. Refrigerate only when ripe, then use as soon as possible. As a rule, pears that mature in the summer should be eaten as soon as they ripen, while autumn and winter pears are firmer fleshed and will keep longer.

Preparation

- Ripe, juicy pears are best eaten fresh, though they are also tasty prepared in a great variety of ways. For cooking, it is best to use slightly underripe fruit. Overripe pears can be made into pearsauce (like applesauce), pear butter, chutney, sorbet and juice.

- Pears are always more aromatic, refreshing and flavourful when chilled only lightly. Once a pear has been sliced, use it immediately or sprinkle with lemon juice to prevent it from discolouring.

Suggestions

- Ripe pears are very good with prosciutto or Parma ham as an hors d'oeuvre or served for dessert with a variety of cheeses, such as cheddar, Camembert, blue cheese, goat cheese, Gorgonzola and Stilton.

- Use pears in any apple recipe. Pears are especially good in pies, mixed with one or two quince. In Europe they are often poached in wine.

- Try pear and ginger chutney or whole pickled seckel pears to compliment a variety of poultry and meat dishes.

Nutritional Value

Bartlett pear

(per 1 medium = 169 g = 6 oz)

			RDA**	%
calories*	(kcal)	100		
protein*	(g)			
carbohydrates*	(g)	26		
fat*	(g)			
fibre	(g)	4.7	30	15.7
vitamin A	(IU)	33	3300	1.0
vitamin C	(mg)	7	60	11.7
thiamin	(mg)	0.03	0.8	3.8
riboflavin	(mg)	0.07	1.0	7.0
niacin	(mg)	0.3	14.4	2.1
sodium	(mg)		2500	
potassium	(mg)	211	5000	4.2
calcium	(mg)	19	900	2.1
phosphorus	(mg)	18	900	2
magnesium	(mg)	9	250	3.6
iron	(mg)	0.4	14	2.9
water	(g)	84		

*RDA variable according to individual needs
**RDA recommended dietary allowance (average daily amount for a normal adult)

- Ripe pears are extremely good in salads, especially with sweet onions and slightly bitter greens, including endives, dandelions, arugula, radicchio and radish cress. Use a tangy vinaigrette or a creamy Roquefort dressing.

- Bake or broil pear halves with poultry, veal, pork or lamb chops, filling the hollows with a little vinegary mint sauce.

- For a delicious dessert, halve or slice 2 or 3 pears, place in a baking dish and sprinkle with lemon juice. Mix 30 mL (2 TBSP) butter, 45 mL (3 TBSP) brown sugar and 30 to 45 mL (2 to 3 TBSP) flour until crumbly. Add chopped walnuts or pecans and a little cinnamon and cloves. Sprinkle topping over pears and bake for about 30 to 40 minutes in a moderately high oven.

- For a simple and pleasing dessert, serve sliced pears with chocolate mint ice cream.

Passe-Crassane ↓

Chile Red Bartlett ↑

Pear and watercress soup

(serves 4 to 6)

Ingredients		
6-8	ripe pears (Bartlett or Anjou)	
1 L	water	4 cups
1	large leek	
30 mL	butter	2 TBSP
	white pepper and salt	
1	chicken or vegetable stock cube	
1	bunch watercress	

Instructions

• Wash, peel, quarter and core pears, reserving the peel and cores. Put the pieces of pear into a bowl of cold water until you are ready for them. Put the peel and cores into a small pot with 2 cups of water; bring to a boil, then simmer for about 20 minutes. Strain, mashing the contents through a sieve, and reserve.

• Clean the leek and slice. In a large pot, melt the butter, add the leek and salt and pepper, then sauté on low heat, covered, for 10 to 15 minutes, until the leek is soft but not browned.

• Add the reserved pear liquid, 2 cups of water, the pieces of pear and the stock cube. Heat, then simmer about 5 minutes.

• Wash the watercress thoroughly in cold water. Pick out each piece, discarding a tiny bit from the end of the stem.

• Add the watercress to the pot and stir once, then quickly remove the pot from the stove, or the cress will lose its fresh colour.

• In a blender, puré until smooth, then adjust the seasoning.

• This soup is excellent cold and can be served hot as well. It develops a better taste when prepared the day before and refrigerated.

Apple pear

↑ Nijisseiki (20th century)

GOOD SOURCE OF POTASSIUM
AND B VITAMINS
LOW IN CALORIES AND SODIUM
SOURCE OF FIBRE

Origin

• Asian pears, commonly called apple pears, are sometimes known as sand pears, oriental pears or shalea (sha li means sand pear in Chinese). They are **not** the result of crossing a pear with an apple, which has never successfully been done. Though both belong to the same family, Rosaceæ, they are not of the same genus. Apple pears are very closely related to pears and probably were their ancestor. Originally from Northern Asia, these prized fruits were cultivated long before pears were. They are still pollinated, wrapped individually on the tree and harvested by hand.

• The best loved fruit for thousands of years in the Orient, where they are considered excellent for the digestion, apple pears were brought to California by Chinese and Japanese immigrants. It is only in recent years that they have been commercially cultivated in California and Washington. China, Japan, Korea and Taiwan are all major producer countries. In the last few years, New Zealanders, who call all apple pears "nashi", have exported them to North America in increasing quantities. Brazil has also begun production.

Varieties and description

• There are approximately 1,000 varieties of apple pears, though only a few are grown commercially. Several characteristics are shared by all varieties — a thin, smooth skin; a very crisp and crunchy texture, similar to a hard apple; and remarkably juicy, refreshing, mild and sweet pear-like flesh. Their sizes range from quite small to more than 500 grams (1 pound) each and they vary in fragrance. All varieties are round except for the uncommon, pear-shaped, yellowy green ya li and tu li.

• The most widely known varieties in North America, nijisseiki (20th century) and shinseiki (new century), have yellowy-green glossy skin. Hosui, kosui and shinsui have golden brown, russeted skin and a more pronounced flavour. Shinsui and the less common chojuro, also russeted, have a more granular texture than the others. Kikusui has a green skin.

Shinseiki (new century) ↓

Availability

• They are available year long because of the large number of producer countries and their ability to stand up well to long storage. American and Japanese apple pears are harvested from August to December, while those from New Zealand are picked from January to April. The season in Taiwan is from late June to mid October.

Selection

• Choose firm, fresh-looking apple pears. Unlike pears, which are always picked green and ripened afterward, apple pears are ripened on the tree and can be eaten immediately after being harvested. Their thin, fragile skin shows scars and marks if handled carelessly. Luckily, these marks are only skin-deep and don't affect the flesh underneath. Most growers use protective packaging to preserve a good appearance.

Storage

• Leave at room temperature if they are to be eaten within a week. Apple pears will keep well for about 2 months when refrigerated, if they are in good condition. Wrap them individually in paper towels or tissue paper so they don't touch each other and place them in a plastic bag in the crisper. Sweet, yellow-skinned varieties will not keep as long as green-skinned ones, while those that are russet-skinned have the longest shelf life.

Preparation

• Apple pears are usually washed and enjoyed in their natural state, alone or in a salad or cold platter. Their light, delicate flavour, best when slightly chilled, can easily be overwhelmed by strong flavours. Their high water content is not always ideal for cooking, though they are tasty when baked or poached. Freezing is not recommended.

• Most people prefer to peel apple pears and slice the fruit crosswise, which shows off a decorative star-shaped centre. Use them as a garnish for a variety of hot or cold meals.

Suggestions

• Apple pears are perfect for fruit salad, as they hold their shape and provide a wonderful texture and flavour. They are also very good in cottage cheese salad or served with yogurt dressing.

• For an extra touch or juicy, crisp texture, add chunks of apple pears to a stir-fry or use in a sweet and sour dish.

• Apple pears make an excellent juice.

↓ Hosui

Shinsui ↓

Apple pear chocolate fondue

(serves 4)

Ingredients

2-4	apple pears	
1/2	lemon	
1	large Toblerone chocolate bar	
30 mL	Poire Williams or Grand Marnier	2 TBSP
75 mL	light cream	1/3 cup
	a few pieces of candied ginger	

Instructions

- Peel 2 large or 4 small apple pears. Divide each apple pear crosswise in 3 pieces. Cut each of these rounds into 6 wedge-shaped pieces, then cut away the core and seeds. Squeeze lemon over the fruit to prevent discolouring.
- In a double boiler, melt the chocolate slowly on low heat with the liqueur. When it is hot, stir in the cream, then remove from heat and transfer to a fondue pot, or a well heated oven-proof bowl.
- Arrange the fruit with slivers of ginger on a platter and serve immediately, providing fondue forks for your guests to dip the fruit and ginger into the chocolate fondue.

Nutritional Value

Apple pear

(per 100 g = 3 1/2 oz)

			RDA**	%
calories*	(kcal)	44		
protein*	(g)	0.4		
carbohydrates*	(g)	11.4		
fat*	(g)	0.2		
fibre	(g)	1.0	30	3.3
vitamin A	(IU)	1.5	3300	0.5
vitamin C	(mg)	4	60	6.7
thiamin	(mg)	.03	0.8	3.8
riboflavin	(mg)	.03	1.0	3.0
niacin	(mg)	0.1	14.4	0.7
sodium	(mg)	7	2500	0.2
potassium	(mg)	141	5000	2.8
calcium	(mg)	10	900	1.1
phosphorus	(mg)	15	900	1.7
magnesium	(mg)		250	
iron	(mg)	0.6	14	4.3
water	(g)	87.6		

*RDA variable according to individual needs

**RDA recommended dietary allowance (average daily amount for a normal adult)

New Zealand apples

(Braeburn/Gala/Royal Gala)

1. Braeburn 2. Royal Gala

VERY GOOD SOURCE OF FIBRE
SOURCE OF IRON AND POTASSIUM
LOW IN SODIUM AND CALORIES

Origin

• The apple, a member of the rose family, Rosaceæ, is one the most esteemed fruits in the Northern Hemisphere. Though many think it originated near the Caspian Sea, it could well have come from Northern Europe. Known to man since prehistoric times, apples were cultivated and prized by ancient civilizations in Egypt, Greece and Rome.

• Apples have long been considered excellent for a healthy diet. They are a good natural laxative, blood purifier, a breath freshener and tooth cleanser and have a beneficial effect on bones, the circulatory system and most of the digestive organs.

• Today, thousands of varieties of apples are grown, though few are mass marketed. Early New Zealand settlers brought apple seeds with them and have been improving and developing new varieties ever since, including Braeburn, Gala and Royal Gala, concentrating on excellence of flavour as much as on good appearance. New Zealand is noted for having introduced the Granny Smith apple to North America about 30 years ago.

• In recent years, Washington and British Columbia have begun cultivation of Gala apples and they are gaining popularity in North America.

Varieties and description

• Developed in New Zealand, the Gala and the more fragrant Royal Gala apples are the result of crossing a superb British apple called Cox's Orange Pippin with Red Delicious and Golden Delicious apples, which were originally developed in Iowa toward the end of the 19th century. Both have thin, creamy yellow skin; the Gala is striped with pink while the Royal Gala has a crimson red blush. Their crisp flesh is sweet and juicy.

• The Braeburn, another variety developed in New Zealand, has been commercially available only since 1985. Its greenish-gold skin is striped with red. It is firm, yet very juicy and sweet, yet tangy.

Availability

• Gala and Royal Gala apples are available from April to June, while Braeburns can be found from May to August. Galas from Washington and British Columbia are available from September to December.

Selection

• New Zealand apples are shipped to market fresh from harvesting. Packaged with great care, they are most often found in prime condition — smooth, firm, brightly coloured and unblemished.

Storage

• Keep apples refrigerated in a plastic bag for several weeks.

Preparation

• Though New Zealand apples are not waxed, they should still be washed before being eaten. Though outstanding fresh, as a snack or in a salad, they are also tasty when cooked.

Suggestions

• Braeburn apples are excellent with cheese.

• For an instant applesauce, use any of these varieties, diced and microwaved for about 2 minutes, covered — delicious served with roasted meats or chops, or enjoyed as a low-calorie dessert.

• New Zealand apples are good in cole slaw, rice salad, chicken or turkey salad and even in tuna or salmon salad.

• For a tasty salad, toss finely sliced red cabbage, a grated carrot, diced New Zealand apples and minced sweet onion with a tasty vinaigrette. A finely sliced fresh fennel adds a gourmet touch.

• For a savoury vegetable side dish, sauté onion and red cabbage with diced New Zealand apples, a few raisins, a pinch of nutmeg, cloves, sugar, salt, pepper and a squeeze of lemon or a touch of vinegar. A few caraway seeds can be added, if desired.

• These apples are excellent in apple and rhubarb compote.

• Dice an apple into a vinegary mint sauce and serve with lamb dishes.

Nutritional Value

Apple

(per 100 g = 3 1/2 oz)

			RDA**	%
calories*	(kcal)	51		
protein*	(g)	0.4		
carbohydrates*	(g)	13.2		
fat*	(g)	0.3		
fibre	(g)	0.6	30	2
vitamin A	(IU)	67.2	3300	2.1
vitamin C	(mg)	4	60	6.8
thiamin	(mg)	.03	0.8	3.8
riboflavin	(mg)	.03	1.0	3
niacin	(mg)	.25	14.4	1.7
sodium	(mg)	2	2500	.08
potassium	(mg)	130	5000	2.6
calcium	(mg)	10	900	1.1
phosphorus	(mg)	10	900	1.1
magnesium	(mg)		250	
iron	(mg)	0.5	14	3.6
water	(g)	86		

*RDA variable according to individual needs
**RDA recommended dietary allowance (average daily amount for a normal adult)

Waldorf fruit salad

(serves 4)

Ingredients		
3	New Zealand apples of your choice	
4	Seckel pears	
3	celery sticks	
1	Boston lettuce	
125 mL	walnuts	1/2 cup
60 mL	raisins	1/4 cup
	Dressing	
60 mL	cream cheese or creamy cottage cheese	4 TBSP
30-45 mL	raspberry or fruit vinegar	2-3 TBSP
30 mL	sunflower oil	2 TBSP
5 mL	fresh mint, finely chopped	1 tsp
1	garlic clove, finely minced	
	salt and pepper to taste	

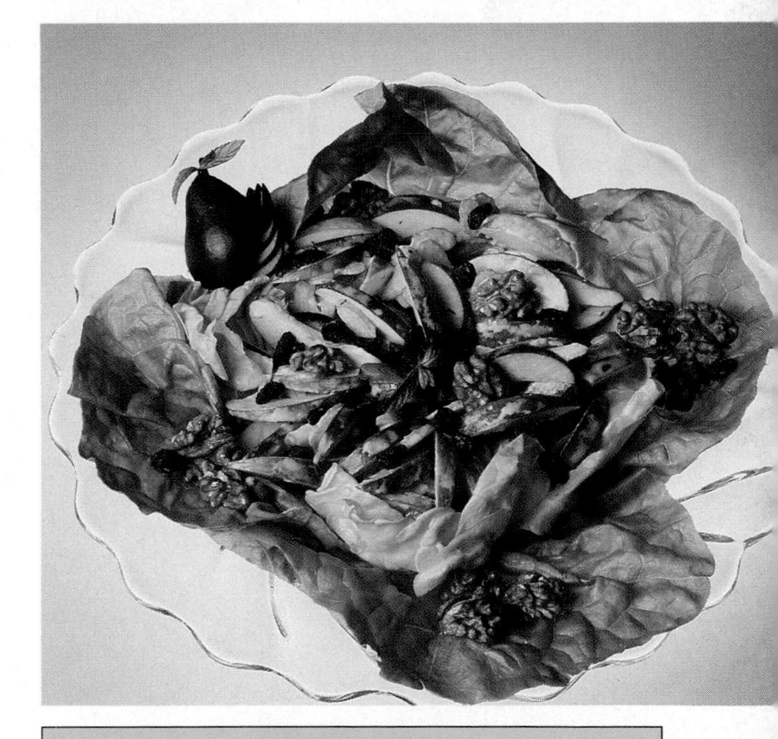

Instructions

• Halve and core apples and pears, but don't peel them. Cut in slices.

• Wash celery sticks and slice finely.

• Blend dressing ingredients until smooth, and toss it immediately with the fruit and celery to prevent browning.

• Arrange the fruit and celery attractively on a bed of lettuce then sprinkle nuts and raisins over it.

Coconut

Source of minerals
Low in fibre

Origin

• The coconut (along with the date), is the most indispensable member of the Palmaceæ family. Of utmost importance to hundreds of millions of people in tropical areas around the world, the coconut palm has provided food, drink, oil, sugar, fuel, housing, furnishing and sometimes even clothing materials for thousands of years.

• Coconuts probably originated in Southern Asia. They are able to float for weeks at sea, wash up on a tropical shore, then establish themselves and proliferate. In this way, they spread to islands throughout the Indian Ocean to Africa and throughout the Pacific islands to the western coast of the Americas, from where they were later introduced to the Caribbean and the Atlantic coast of the Americas in the 16th century. Today they grow in warm areas throughout the world, flourishing best near the equator.

Varieties and description

• A coconut is a very large, green, elongated fruit. It takes about a year to mature and can be enjoyed at several stages of its development. At first, only the refreshing liquid is drunk. At about 6 months, young coconuts contain liquid and a thin, gelatinous coating of extremely nutritious, tasty, white flesh that is simply eaten with a spoon or used in tropical cuisine.

• The mature coconuts that are exported have had their outer green husks removed. Usually 10 to 15 centimetres (4 to 6 inches) in diametre, they are round, hard-shelled, dark brown fruits covered with fibres. There are 3 "eyes", or indentations, fairly close together. Inside the shell, the part that is eaten is actually a seed, and consists of a fibrous, sweet, nutty-tasting white flesh, about 1-centimetre (1/2-inch) thick, covered by a thin edible brown layer. The nutritious colourless coconut water in the centre is gradually absorbed into the flesh as the coconut matures.

• The exotic Southeast Asian **salak,** related to the coconut, is also a member of the palm family. The size of a small plum, the salak has an extraordinary, thin, glossy, reddish brown, scaly skin that looks like fine alligator or snake skin. The firm, dryish, ivory-coloured flesh is divided into three segments, each enclosing a shiny brown seed.

• Another well-known delicacy from the palm family, the hearts of very young palm shoots, are long, white, cylindrical and have a crunchy texture. Their bland flavour is complimented by a spicy vinaigrette.

Availability

• Fresh coconuts can be found year round. They are most plentiful from October to January.

• Salaks, available from May to August, are only rarely imported.

Selection

• Look for heavy coconuts that contain liquid. Shake them to hear it slosh around. Make sure there is no sign of moisture near the eyes, or any smell of fermentation.

Storage

• Keep coconuts at room temperature and use within a week, or refrigerate for several weeks. Freeze leftover grated coconut, or dry it out in a slow oven and store in an airtight container. Once opened, a coconut can be wrapped and refrigerated for a few days.

Preparation

• Make a hole in 1 or 2 of the eyes, using a sharp pointed object, then drain out the coconut water. Drink it

Salak ↓

chilled, alone or mixed with juice. Place the whole coconut in the oven for about 15 minutes, at 130 ° to 160 °C (250 ° to 325 °F) to help the flesh shrink away from the shell. Tap the coconut with a hammer to break the shell and remove the flesh. If the flesh is still difficult to remove, return the coconut to the oven for another 5 or 10 minutes. The thin brown skin can be left on or peeled off, as desired.

• While North Americans most often enjoy their coconuts either freshly grated or cut in small pieces, coconut milk is used far more often in tropical cuisine. To make coconut milk, grate the flesh by hand or in a food processor. Combine it with 3 or 4 cups of water, boil, then simmer for a few minutes, stirring constantly. Let cool, then strain through cheesecloth, squeezing out as much liquid as possible. Afterward, discard the solid part. Coconut milk will keep for only a few days in the refrigerator, but can be frozen for longer storage.

Suggestions

• Grate fresh coconut and sprinkle over fruit salad, ice cream, cakes, cookies, pies and pastries, or add to cereal.

• Rice and rice pudding are delicious cooked in coconut milk.

• Use coconut milk in sauces, soups or stews.

• Shredded coconut and coconut milk are excellent in curries.

• Combine cold coconut milk and rum to make an exotic tropical drink called coco loco.

Nutritional Value
Fresh coconut
(per 48 g, shredded = 1 3/4 oz)

			RDA**	%
calories*	(kcal)	174		
protein*	(g)	1.6		
carbohydrates*	(g)	6.8		
fat*	(g)	16.3		
fibre	(g)		30	
vitamin A	(IU)		3300	
vitamin C	(mg)	2	60	3.3
thiamin	(mg)	0.02	0.8	2.5
riboflavin	(mg)	0.01	1.0	1.0
niacin	(mg)	0.3	14.4	2.1
sodium	(mg)	8	2500	0.3
potassium	(mg)	373	5000	7.5
calcium	(mg)	10	900	1.1
phosphorus	(mg)	47	900	5.2
magnesium	(mg)		250	
iron	(mg)	0.95	14	6.8
water	(g)	24.4		

*RDA variable according to individual needs
**RDA recommended dietary allowance (average daily amount for a normal adult)

Coconut crisps

Ingredients	
1	coconut

Instructions

• Open the coconut and use a very sharp knife or a vegetable peeler to shave off thin strips or curls of coconut flesh.

• Spread the pieces out on baking sheets and bake at 130 °C (250 °F) until golden brown, from 1 to 2 hours, turning once or twice. Sprinkle on salt if desired.

• Cool and serve as a snack. Store leftovers in an airtight container and crisp in the oven again, if necessary, before serving.

Specialty bananas

Origin

• Bananas belong to the Musaceæ family and it may surprise most people who are only familiar with the well-known Cavendish type that there are about 300 varieties of banana around the world, most of which are known only to their native countries. While the fruit has long been a food staple in the tropics, the plant is extremely useful for thatching and other purposes.

• Most botanists believe that bananas originated in Southeast Asia and that they were one of the first crops cultivated by man. (Wild bananas are actually quite unpalatable and full of seeds.) They eventually spread west to India a few thousand years ago, where Alexander the Great saw them in 327 BC; later, they were introduced to Africa, possibly by Arab traders. Fra Tomás de Berlanga is credited with introducing them to the Caribbean Spanish colony on Hispaniola in 1516, where they thrived and from where they were disseminated throughout the tropical areas of the New World. It is possible the banana was already established on the Pacific coast at that time, brought by pre-Columbian Indians who migrated from Southeast Asia.

• Bananas, called "the food of the wise", are excellent for low-sodium, low-fat and low-cholesterol diets. They are good for those suffering from ulcers or colitis. A good natural laxative, bananas are beneficial for the very young and the very old. They are often prescribed for infant diarrhea and any seasoned traveller with stomach problems will tell you that nothing goes down as well as a banana with a squeeze of lemon or lime juice!

• Though bananas were sporadically imported in the late 19th century as a tropical luxury for the rich, it was not until the 1920s, when proper transportation and refrigeration started to become available, that bananas became well known in North America and Europe and available to everyone. Today, most bananas imported into North America are grown in Central and South America, the Caribbean and Mexico, though small crops are raised in California and Florida.

Varieties and description

• Popular among the new specialty bananas, are the small sweet, yellow varieties, which are often referred to as dwarf, **baby** or **ladyfinger** bananas. Some are shaped like miniature slender bananas, such as the small **orito** banana, which has a hint of pear flavour. Other types include the stocky apple banana **(manzano),** a delicious banana with a suggestion of apple and strawberry flavour. The chunky **burro** banana from Mexico, which has a slight lemony flavour and the **blue Java** or **ice cream** banana are also gaining in popularity.

• The plump pink, red or purple-skinned bananas **(colorados),** creamy-coloured and tinged with pink or orange on the inside, are usually sweeter than yellow

DWARF BANANAS ARE AN EXCELLENT SOURCE OF POTASSIUM AND IRON
VERY LOW IN FIBRE
LOW IN SODIUM

Plantain ↑

Baby or ladyfinger bananas ↓

bananas. These come in a variety of sizes, from mini to normal, though most are shorter than **Cavendish** bananas.

• The large, thick-skinned plantain — longer, much firmer, starchier and less sweet than other types of banana — is always cooked. It should be treated more like a vegetable than like a fruit.

Availability

• Although bananas grow year round, unusual varieties are imported only sporadically.

Selection

• Bananas are picked green, transported to their destination and then ripened. The riper they are, the more easily digestible and sweet they become. Do not worry if bananas have brown or black speckles on their skin, as this is only an indication of their ripeness. Avoid bananas that have a split skin.

• If green or yellow plantains are selected, they will need to be ripened. The tastiest plantains have nearly black skin.

Storage

• Keep bananas at room temperature until the yellow varieties begin to show dark spots and the red varieties turn a bit darker. When they are very ripe they can be refrigerated for a few days. Apple bananas (manzanos) taste best when their skin has turned black. Plantains should not be refrigerated unless they are completely ripe.

↓ **Red-skinned (colorado)**

Preparation

• Specialty bananas are most often enjoyed when fresh, though the small size of baby bananas can be useful and decorative for certain dishes. Along with red bananas, they look very appealing in a fruit bowl.

• Plantains must be cooked to be palatable. Either stewed, baked, grilled or fried, they usually accompany roast meats, beans and rice. Be sure to discard all the strings that run its length just under the skin. When still green or partially ripe, its taste is more like squash or potato. When the skin is black, it tastes more like banana or sweet potato.

Suggestions

• Bananas are amazingly versatile. Try a refreshing banana shake or an exotic banana daiquiri. Baby bananas are ideal for banana splits.

• Bananas are used in West Indian, Latin American, African and Asian cuisine. They are very good in curries.

• For a Caribbean dessert of baked bananas, slice them in half or thirds lengthwise, sprinkle on cinnamon, brown sugar, lime juice, a little butter, and a bit of rum. Bake in a hot oven about 20 to 25 minutes. Sliced oranges can be added.

• For tasty Caribbean banana fritters served with the main course or with maple syrup for dessert: Mash 2 ripe bananas, beat in 1 egg and combine with 125 mL (1/2 cup) milk, 125 mL (1/2 cup) flour, 5 mL (1 tsp) baking powder, 15 mL (1 TBSP) sugar. Fry like small pancakes.

• To make banana chips: Add 1 to 2 centimetres (1/2 inch) oil to a fry pan, heat to 190 °C (375 °F), slice bananas thinly (green bananas or partially ripened plantains are best) and fry until brown and crisp. Drain on paper towels and salt lightly, if desired.

• Bake plantains in their skins: First, slit skin lengthwise and place slit side up, then bake at 180 °C (350 °F) for about 45 minutes to an hour. When done, discard skins and season with butter and brown sugar.

• To fry plantains, peel and cut in thirds lengthwise, dust with flour and cinnamon, then fry in margarine or butter until browned.

Flambéed baby bananas

(serves 4)

Ingredients		
4-6	baby bananas	
45 mL	butter	3 TBSP
45 mL	brown sugar	3 TBSP
10 mL	lime or orange zest	2 tsp
	juice of 1 lime	
60 mL	rum or orange liqueur	1/4 cup
4-8	thin orange slices, halved	
4-8	thin lime slices, halved	
	cinnamon to taste	
30 mL	cognac	2 TBSP

Instructions

- Cut bananas in half lengthwise or leave whole.
- Melt the butter in a large frying pan, then add sugar, finely grated citrus zest and the lime juice.
- Sauté the bananas about 5 minutes, turning a few times. Remove to a hot serving platter.
- Add the rum to the frying pan, then the slices of citrus fruit. Sauté on high heat for a minute or two, then arrange the fruit around the bananas, drizzling the rest of the sauce over the bananas and sprinkling on cinnamon.
- Heat the cognac and pour it over the fruit at the table, setting it alight immediately.
- Serve with ice cream or whipped cream, if desired, and coffee.

Nutritional Value

Dwarf banana

(per 100 g = 3$^{1}/_{2}$ oz)

			RDA**	%
calories*	(kcal)	72		
protein*	(g)	1.8		
carbohydrates*	(g)	18		
fat*	(g)	0.2		
fibre	(g)	0.2	30	0.7
vitamin A	(IU)	266	3300	8
vitamin C	(mg)	8	60	13
thiamin	(mg)	0.03	0.8	3.8
riboflavin	(mg)	0.04	1.0	4
niacin	(mg)	0.6	14.4	4
sodium	(mg)	48	2500	1.9
potassium	(mg)	435	5000	8.7
calcium	(mg)	10	900	1.1
phosphorus	(mg)	24	900	2.7
magnesium	(mg)		250	
iron	(mg)	1.3	14	9.3
water	(g)	80		

*RDA variable according to individual needs

**RDA recommended dietary allowance (average daily amount for a normal adult)

Champagne grapes

Origin

• These highly decorative, sweet and delicious miniature currant grapes are actually a variety called Zante currant or Black Corinth, named for the Greek city where they were prized more than 2,000 years ago both fresh (when in season) and dried. From the Vitaceæ or vine family, Zante currants are not related to redcurrants or blackcurrants, which are berries of the Saxifragaceæ family.

• Grapes are thought to have originated in the region of the Caspian Sea. The ancient Greeks were the first Europeans to cultivate them and the champagne grape variety is believed to date back to that era.

• Brought to California by Greek immigrants nearly 100 years ago, today they are a thriving crop in California's San Joachim Valley. Formerly, the entire crop was dried and used for baking, but when mini-vegetables and exotics came into fashion in the early 1980s, they were introduced to the foodservice industry as a gourmet garnish. Small bunches were draped over champagne or wine glasses, hence their popular name, champagne grapes. Their wonderful, sweet flavour and attractive, unique appearance has opened up a new market for them as a dessert fruit.

Description

• Champagne grapes are very small and seedless, with a deep blue-black colour at maturity. The average size for each bunch is the length and width of a hand.

Availability

• California champagne grapes are one of the earliest maturing of all grape varieties. They are usually found from July to October.

Selection

• Grapes are ripened on the vine and harvested only when they have reached their peak of perfection. Choose unblemished champagne grapes with fresh, pliable stems.

Storage

• Champagne grapes should be stored in the refrigerator, preferably wrapped in tissue or a paper towel, then placed in a perforated plastic bag. This should keep them from dehydrating. Though they will keep well for a week or two, they are best eaten as soon as possible.

Preparation

• Wash and drain well, then enjoy these delectable little fruits just as they are. Each bunch has a main stem that runs its length. Use a scissors to snip off small clusters where they join the main stem. Hold the tip of the finer sub-stem of each small cluster and simply pop the bunch into your mouth; as you pull gently, all the fruit will pull free. Any tiny bits of the soft stem that might remain attached to the fruit are edible.

Suggestions

• Champagne grapes have an exceptionally high sugar content, enabling them to be frozen for later use on special occasions. Its best to cut them into small bunches first and spread them out on a plate or tray. When they are fully frozen, put them in an airtight container and return them to the freezer.

• Make a beautiful fruit platter by arranging sliced mangos or melons in the centre and surrounding them with a ring of small clusters of champagne grapes.

• For a touch of elegance, use a small cluster to garnish any appetizer, main course, drink or dessert.

• Use champagne grapes as a breakfast food, snack or lunchbox treat.

Celeriac

VERY LOW IN CALORIES AND SODIUM
SOURCE OF VITAMIN C
AND PHOSPHOROUS
LOW IN FIBRES

Origin

• Celeriac, sometimes called turnip root celery, celery knob or celery root, is not the root of common celery but another variety cultivated for its enlarged root. It originated in the Mediterranean region and was well known to the ancient Greeks and Romans, who believed it to be a good blood purifier. A member of the Umbellifer family, celeriac is related to carrots, parsley, fennel, dill, chervil and, of course, celery. It has been cultivated far longer than celery, and is, in fact, far easier to grow. A popular crop in many European countries for hundreds of years, celeriac was better known in North America during the last century than it is today, though it is still grown here.

Varieties and description

• Celeriac is a large, unusual-looking edible root, very irregular in shape, with clumps of rootlets sticking out. In size it is comparable to a large turnip. Its skin is brown and rough and its flesh is creamy white.

• Its spindly stalks and leaves contain aromatic oils and smell strongly of celery. Its firm flesh tastes crisp and nutty.

• In Europe, hybridized varieties have more regular shapes.

Availability

• Celeriac can be found year round, though it is most plentiful from fall until early spring.

Selection

• It is best to select a medium-sized, very firm celeriac, about 400 to 500 grams (1 pound). The less irregular the shape, the easier it is to peel and the less you will waste. If you do encounter a soft spot inside, simply cut it out and use the rest.

• Very large celeriacs may be woody-textured.

Storage

• If the stalks and tops are still attached, cut them off. (Try using them to give a strong celery flavour to soup stocks.) Keep the root in a plastic bag in the refrigerator for up to several weeks.

Preparation

• Wash well, then peel. Keep peeled celeriac in water with a touch of lemon or vinegar to prevent it from turning brown. If you are grating it for salad, it is a good idea to sprinkle on lemon juice or vinegar as you grate it.

• Celeriac has the best flavour when raw, but is often cooked in Europe. It is possible to cook it first and peel it afterward, although it needs to be peeled and chopped first for certain dishes. Be careful not to overcook celeriac, as it will lose flavour and become mushy in texture.

Suggestions

• For the simplest version of the classic French dish "céleri-rave rémoulade": grate celeriac and add white wine vinegar or lemon juice, olive oil, Dijon mustard, mayonnaise, salt and white pepper. A touch of celery salt adds a good flavour. Keep leftovers for several days in the refrigerator.

• Grated celeriac is delicious in salmon, tuna or egg salad.

• Use celeriac, lightly cooked, in any puréed soup to add flavour. Dice it and add to hearty vegetable or bean soups or to stews.

• To prepare as a side vegetable, dice and boil celeriac for 5 to 10 minutes, flavouring with butter, salt and herbs.

• Cube or julienne, cook lightly and prepare it au gratin with a little mornay or béchamel sauce. Sautéed onions are excellent included in this dish.

Nutritional Value Celeriac (per 100 g = 3½ oz)				
			RDA**	%
calories*	(kcal)	40		
protein*	(g)	1.8		
carbohydrates*	(g)	8.5		
fat*	(g)	0.3		
fiber	(g)	1.3	30	4.3
vitamin A	(IU)		3300	
vitamin C	(mg)	8	60	13.3
thiamin	(mg)	0.05	0.8	6.3
riboflavin	(mg)	0.6	1.0	6.0
niacin	(mg)	0.7	14.4	4.9
sodium	(mg)	100	2500	4.0
potassium	(mg)	300	5000	6.0
calcium	(mg)	43	900	4.7
phosphorus	(mg)	115	900	12.8
magnesium	(mg)		250	
iron	(mg)	0.6	14	4.3
water	(g)	88.4		

*RDA variable according to individual needs
**RDA: recommended dietary allowance (average daily amount for a normal adult)

Celeriac and carrot salad

(serves 4)

Ingredients		
1	medium celeriac	
1-2	carrots	
2-4	scallions (green onions), minced	
30 mL	fresh lemon juice	2 TBSP
45-60 mL	mayonnaise	3-4 TBSP
1	garlic clove, finely minced	
	salt and pepper to taste	

Instructions

- Peel and grate celeriac and carrots.
- Mix vegetables in a salad bowl, add dressing and mix well. Refrigerate for an hour.
- Serve as is, or on a bed of lettuce. Garnish with carrot slivers.

Celeriac and carrot salad →

Celeriac slaw

(serves 6-8)

Ingredients		
1	celeriac	
3-4	carrots	
1	daikon	
1	yellow or orange pepper	
1	green apple	
1	small sweet onion	
15-30 mL	fresh dill, finely minced	1-2 TBSP
30 mL	parsley, finely minced	2 TBSP
60 mL	mayonnaise	4 TBSP
30 mL	sunflower seed oil	2 TBSP
	juice of 2 lemons	
5-10 mL	white horseradish sauce	1-2 tsp
	salt and freshly ground black pepper to taste	

Instructions

- Peel and grate the celeriac, carrots and daikon (a long white oriental radish).
- Cut pepper in fine slivers, dice apple and finely mince onion.
- Mix the rest of the ingredients together to make the dressing.
- Toss the dressing and vegetables thoroughly and chill until ready to serve.

Salsify

Very good source of iron and other minerals
Contains vitamin C
Low in calories when harvested

Origin

• White salsify — also known as vegetable oyster, oyster plant or oatroot — and the closely related black salsify, viper grass or scorzonera, originated in the Mediterranean area. Scorzonera derives its name from the Catalan word "escorso", meaning viper, as centuries ago its milky sap was thought to be an effective treatment for snakebite (or perhaps because it looks like a black viper). Both plants are members of the large Compositæ family, which includes lettuce, artichokes, endives, Jerusalem artichokes and burdock or gobo, a popular root in Japan and Asia that resembles black salsify.

• Though wild salsify was known to man for more than 2,000 years, it was not cultivated in Southern Europe until the 17th century. Popular in England and France, it was brought to the New World by early colonists. Although it was never widely grown, it was planted in most vegetable gardens. Over the years, it has been cultivated only in small quantities in North America, but it is well liked in Europe, especially in France and Belgium. Belgium is one of the world's largest producers; it is also grown in France, Holland and Italy.

Varieties and description

• White salsify resembles parsnip with its pale, brownish skin, though parsnip has a single taproot and salsify usually has branching or forked roots. Its fresh leaves, called "chards", can be cooked like spinach or used in salad; they taste somewhat like endive.

• Black salsify is longer, more slender and doesn't branch. It has a dark brown to black skin. Its creamy coloured flesh has a stronger flavor than that of white salsify.

• Both kinds of salsify are said to have a delicate oyster-like taste, with hints of artichoke or asparagus. Some even find an aftertaste of coconut! Like Jerusalem artichokes, it contains inulin, which may give some people gas if eaten in large quantities.

Availability

- Salsify can be found from September to May.

Selection

- Choose medium-sized, fresh, firm roots. The fresher the salsify, the better the flavour. Like parsnips, salsify is sweeter if exposed to cold temperatures before harvesting.

Storage

- Refrigerate unwashed salsify in the crisper, wrapped in a plastic bag with several small air holes. It should be used within a few days.
- Salsify should never be frozen, even after preparation.

Preparation

- Salsify must be washed and peeled, but remove only the thinnest layer possible. The flesh of both white and black salsify will begin to darken immediately when exposed to air. Some cooks boil the roots whole for about 15 minutes and then remove the skin, thus avoiding discolouration; or cut the roots in 3 or 4 pieces and after peeling each one, drop it into cold water with vinegar or lemon juice.
- Raw salsify has a very bland and delicate taste. It needs to be cooked until just tender to develop a good flavour, but be careful not to overcook it.
- Peeling black salsify may stain your fingers temporarily. You can avoid this by wearing plastic gloves.

Suggestions

- Salsify is very good when simmered in water for 15 minutes, then sautéed with a little garlic and chopped parsley in a seasoned mixture of oil and margarine or butter for another 10 to 15 minutes, or until the pieces are well browned.
- Boil or steam salsify, then glaze it in butter and a little sugar, browned butter, or butter with lemon and finely chopped parsley.
- It is delicious served au gratin in a light white or cheese sauce. For an appetizing addition, include sautéed onions with a pinch of nutmeg.
- Salsify is excellent combined with vegetables such as onions, potatoes, leeks, celery and spinach in a puréed soup or cut coarsely and added to a hearty winter soup.
- For a tasty dish, try grated salsify mixed with an egg, breadcrumbs and seasoning. Make into little pancakes and fry in a Teflon pan with very little oil, until golden brown. A little minced onion can be added to the mixture if desired. Delicious served with sour cream or yogurt. Mix some lemon juice or vinegar into the salsify as you grate it to prevent it from discolouring.
- Salsify is very pleasant braised, stewed or in a casserole with veal, chicken or fish. If the salsify is precooked, add it during the last few minutes of preparation.
- Salsify can be roasted like parsnips: slice salsify in thin strips lengtwise, place in a shallow baking dish and brush both sides with a very small amount of oil. Sprinkle on a little salt and pepper. Bake in a medium-high oven [approximately 190 °C - 200 °C (375 °F - 400 °F)] for 15 to 20 minutes, until browned. Turn once to brown both sides evenly.
- For a simple dish, steam whole roots for about 15 minutes or until tender being careful not to overcook them. Spread them out in a shallow baking tray, sprinkle on salt, pepper, grated nutmeg and a generous portion of grated gruyère or cheddar cheese. Broil a few minutes until the cheese has melted and browned a little.
- Soak the grated salsify in a bowl of cold water with the juice of 1 lemon. Drain before serving with the following dressing: Whisk together 1 egg yolk, 15 mL (1 TBSP) Dijon mustard, 30 mL (2 TBSP) lemon juice, 15 mL (1 TBSP) fresh finely chopped chervil, salt and pepper. Whisk in 75 mL (5 TBSP) sunflower oil, drop by drop, beating constantly. (See photograph next page.)

Nutritional Value

Salsify

(per 100 g = 3 1/2 oz)

			RDA**	%
calories*	(kcal)	16-94***		
protein*	(g)	2.9		
carbohydrates*	(g)	18		
fat*	(g)	0.6		
fibre	(g)	1.0	30	3.3
vitamin A	(IU)	10	3300	0.3
vitamin C	(mg)	11	60	18.3
thiamin	(mg)	0.04	0.8	5.0
riboflavin	(mg)	0.04	1.0	4.0
niacin	(mg)	0.3	14.4	2.08
sodium	(mg)		2500	
potassium	(mg)	380	5000	7.6
calcium	(mg)	47	900	5.2
phosphorus	(mg)	66	900	7.3
magnesium	(mg)		250	
iron	(mg)	1.5	14	10.7
water	(g)	77.6		

*RDA variable according to individual needs
**RDA recommended dietary allowance (average daily amount for a normal adult)
***Caloric value ranges from 16 Kcal/100 g for freshly harvested to 94 Kcal/100 g for stored salsify.

Salsify with scallops

(serves 4)

Ingredients

500-750 g	salsify	1- 1½ lbs
	lemon juice	
30 mL	butter	2 TBSP
30 mL	vegetable oil	2 TBSP
1	medium onion, finely sliced	
2	carrots, julienned	
1	leek, finely sliced	
1-2	garlic cloves	
1	stock cube, fish or chicken	
15 mL	flour	1 TBSP
125 mL	water	1/2 cup
125 mL	white wine	1/2 cup
750 g	scallops	1½ lbs
2 mL	saffron (optional)	1/2 tsp
125 mL	cream or sour cream	1/2 cup
	salt and pepper to taste	
	handful of parsley	

Instructions

- Wash salsify well, peel, cut into pieces 2 to 5 centimetres (1 to 2 inches) long and place them immediately in a pot with cold water and lemon juice, to prevent them from discolouring.
- Add a little salt, bring to the boil, then simmer for about 10 minutes.
- Drain and keep covered.
- In a large skillet, heat the butter and oil and gently sauté the vegetables and garlic for about 5 minutes. Crumble the stock cube and add it.
- Sprinkle on the flour and stir well. Slowly mix in the wine and water and add the partly cooked salsify. Simmer for about 8 minutes.
- Wash the scallops and dry them well. If they are large, cut them in half. Add them and sauté for about 2 minutes.
- Add the cream and seasoning, then continue cooking for about 2 or 3 minutes, until the scallops are lightly done.
- For a lovely, rich colour, add saffron just before the cream.
- Garnish with finely chopped parsley and serve with Swiss chard or spinach and rice.

Sweet potato

EXCELLENT SOURCE OF VITAMIN A AND IRON
GOOD SOURCE OF CALCIUM AND VITAMIN C
VERY LOW IN SODIUM

Orange flesh ↑

Origin

• The sweet potato, native to the tropics and subtropics of the New World, was introduced to Europe by Christopher Columbus. Surprisingly, the sweet potato is a member of the Convolvulaceæ or morning glory family and is not related to the potato. Nor is the sweet potato the same as a yam, though many American growers mislabel them so. Yams require even hotter growing conditions than are available in the U.S. and are virtually unknown outside their growing areas. They are much starchier and not nearly as tasty as sweet potatoes.

• Sweet potatoes were a staple food in the West Indies, Central and South America, Mexico and the southern regions of the United States long before the Europeans arrived in the New World. Early Spanish explorers spread their cultivation to the Philippines and other Pacific colonies, from where it spread to most Asian countries. It is uncertain when or how it reached Africa.

• Sweet potatoes were one of the most important crops in the first American colonies and remained so until well after the American revolution. Today, most of the sweet potatoes consumed in North America are grown in the Southern United States.

Varieties and description

• There are several hundred varieties of sweet potatoes, but only a few are commercially widespread. Sweet potatoes are usually divided into two categories, moist-fleshed and dry-fleshed. This refers to their taste and texture, not to their water content; the dry-fleshed kind actually has a higher water content. The main difference is that the moist-fleshed varieties convert more of their starches to sugars during the cooking process.

• Sweet potatoes are large tuberous roots, about the size of large baking potatoes, often with tapering ends. Their thin skin can be tan, copper, orange or reddish brown, while their flesh can be white, yellowish or different shades of orange. Different varieties vary in sweetness, a factor that does not depend on their colouring.

Availability

• Though they are harvested from August to November, sweet potatoes can be found most of the year, thanks to a curing process that lengthens their shelf life.

Selection

• Choose very firm, unblemished sweet potatoes.

Storage

• Sweet potatoes will keep for a month or two in a cool, dark, airy place. They should not be refrigerated, as cold temperatures and humidity cause soft spots. Cooked, they will keep for about a week in the refrigerator or much longer in the freezer.

Preparation

• Sweet potatoes are best washed and baked whole in a hot oven for about 45 minutes or until they are tender. If

↓ White flesh

the recipe calls for boiling, leave them whole and unpeeled and cook for about 20 to 30 minutes. If a recipe calls for peeled potatoes, peel, then leave them in cold water until they are needed, to prevent discolouration.

• To cook a medium to large sweet potato in a microwave, pierce it in several places, wrap it in a paper towel, and microwave at high for about 5 to 7 minutes, turning once. (Let stand two minutes before serving.) Steaming is not recommended.

Suggestions

• Baked sweet potatoes are delicious with a little butter and salt.

• Once baked, they can be cut in half and have cinnamon, nutmeg, cloves, ginger, brown sugar, orange juice and zest, or even small chunks of pineapple mashed into their soft flesh, with a little butter or magarine.

• Combine leftover cooked sweet potatoes with a minced onion, a grated apple, a small handful of breadcrumbs and the seasoning of your choice and form small patties. Fry until golden brown in a small amount of margarine or oil in a non-stick frying pan.

• Try baking sweet potatoes, whole or cut in large pieces, with roast beef, turkey, pork or chicken, for about 45 minutes to an hour, turning several times to brown them all over.

• Toasted peanuts or pecans add an extra good touch to candied sweet potatoes, which are a traditional accompaniment to turkey and roast pork.

Nutritional Value

Sweet potato

(per 100 g = 3 1/2 oz)

			RDA**	%
calories*	(kcal)	141		
protein*	(g)	2.1		
carbohydrates*	(g)	32.5		
fat*	(g)	0.5		
fibre	(g)	0.9	30	3
vitamin A	(IU)	8100	3300	245
vitamin C	(mg)	22	60	36.6
thiamin	(mg)	0.09	0.8	11.3
riboflavin	(mg)	0.07	1.0	7
niacin	(mg)	0.7	14.4	4.9
sodium	(mg)	12	2500	0.5
potassium	(mg)	300	5000	6
calcium	(mg)	40	900	4.4
phosphorus	(mg)	58	900	6.4
magnesium	(mg)		250	
iron	(mg)	0.9	14	6.4
water	(g)	63.7		

*RDA variable according to individual needs

**RDA recommended dietary allowance (average daily amount for a normal adult)

Sweet potato casserole

Ingredients		
4-5	sweet potatoes	
30 mL	butter or margarine	2 TBSP
Up to 125 mL	milk	Up to 1/2 cup
	handful of chives or scallions, minced	
2 mL	nutmeg	1/2 tsp
	salt and pepper to taste	
2	eggs, separated	
1	peeled orange	

Instructions

• Boil the sweet potatoes until they are tender, then peel and mash them with butter and enough milk to make a creamy, smooth texture. Mix in seasonings and egg yolks.

• Peel and segment orange, removing membrane. Chop in small pieces and add.

• Beat the egg whites until stiff and carefully fold them in.

• Put in one large or 4 small individual oven-proof dishes, sprinkling the top with slivered almonds and grated gruyère cheese.

• Bake at 160 °C (350 °F) for 15 to 20 minutes and serve as a vegetable side dish.

Plum tomato and sun dried tomato

Plum tomato ↑

Origin

- The tomato, a member of the large Solanaceæ family, originated in the South American Andes region. The Incas gathered them from the wild when they were still a small, yellow, berry-like fruit, probably a type of cherry tomato, thousands of years before man began to cultivate them. Having spread throughout the Latin American areas before the arrival of Europeans, tomatoes were taken back to Europe by early Spanish explorers. They were considered an ornamental plant and it took several centuries before they were accepted as a food.

- The Spanish and later the Italians were the first (and for a long time the only) people to appreciate the versatility of tomatoes. Since the mid-18th century, the Italians have created dozens of varieties, some of which were later taken to North America by settlers. Though Thomas Jefferson tried to interest Americans in tomatoes, they did not really become widely accepted anywhere outside the Mediterranean area until early this century. At one time, cooks recommended boiling them for at least three hours before they were considered edible! Europeans introduced tomatoes to Asia and Africa, where they became an important food.

RICH IN POTASSIUM AND VITAMIN A
GOOD SOURCE OF VITAMIN C AND FIBRES
LOW IN CALORIES

↓ Sun dried tomato

• A superior variety for cooking purposes is the plum tomato, sometimes called a romano or Italian tomato. Today, plum tomatoes are grown all over the Mediterranean area, especially Italy, Mexico, the United States and, during the summer, Canada, among other countries.

• Sun dried tomatoes are a fairly new product, developed for the gourmet food industry. They are mostly processed in California and Italy.

Varieties and description

• Plum tomatoes are small and red, with an oval or pear shape. They contain fewer seeds and a lower water content than other types of tomatoes and so are superb for sauces and stews, though they can also be enjoyed in their natural state.

• Sun dried tomatoes are dark, reddish brown and wrinkled, with a fabulous taste and the chewy texture of dried fruit.

Availability

• Plum tomatoes are available year round, with peak supplies in the summer. Sun dried tomatoes, packaged like other types of dried fruit, can be obtained from specialty shops.

Selection

• Choose firm though not hard, bright red, unblemished plum tomatoes. Softer tomatoes are ripe and should be used immediately.

Storage

• Leave plum tomatoes unrefrigerated unless they are very ripe. Then they should be used within a few days. They can be cooked and frozen for later use.

Preparation

• Slice plum tomatoes into salads or peel them for cooking. Pour boiling water over them and leave for about 30 seconds to a minute and the peel will come off easily, then chop or slice. They are also good sliced lengthwise and grilled or fried with a little olive oil and seasoning.

• Sun dried tomatoes can be soaked in water for a few minutes, then sliced and added to your recipe. They can be added as is to a stew or liquid sauce that requires further cooking.

Suggestions

• For a delicious plum tomato soup, chop 2 or 3 large onions and garlic cloves and sauté in a little olive oil on low heat in a covered soup pot for about 15 minutes. Add a good quantity of peeled, chopped plum tomatoes and a whole bunch of dill, chopped. Continue to cook on low heat, covered, for another 15 minutes. Add 1 to 1.5 litres (4 to 6 cups) water, 2 or 3 stock cubes (vegetable, chicken or fish), a pinch of sugar and salt and pepper to taste. Boil, simmer for 30 minutes, then blend until smooth. Add a little sherry for a richer flavour. Cooked rice or pasta can be added if desired and a little fresh dill makes a nice garnish.

• Steep sun dried tomatoes in water for about 15 minutes, dry thoroughly, then place them in a small jar filled with olive oil, making sure that the tomatoes are completely covered or they will not last long. For extra flavour you can add herbs — bay leaf, garlic, thyme, basil, dill, etc. Use the tomatoes for risotto or pasta sauces, or for a delicious antipasto appetizer along with fresh mozzarella, grilled eggplant slices, roasted peppers and olives, then use the oil for salad dressings.

• Cut sun dried tomatoes into small pieces and marinate them in vinaigrette, then add to a salad.

• Slice plum tomatoes lengthwise, cut thin slices of feta or mozzarella cheese and sprinkle with fresh basil leaves or finely chopped chives. Add a drizzle of olive oil, vinegar, salt and pepper.

Nutritional Value

Plum tomato

(per 100 g = 3 1/2 oz)

			RDA**	%
calories*	(kcal)	20		
protein*	(g)	1.2		
carbohydrates*	(g)	4.2		
fat*	(g)	0.3		
fibre	(g)	0.7	30	2.3
vitamin A	(IU)	1682	3300	51
vitamin C	(mg)	23	60	38
thiamin	(mg)	0.06	0.8	7.5
riboflavin	(mg)	0.04	1.0	4
niacin	(mg)	0.4	14.4	4.2
sodium	(mg)	4	2500	0.2
potassium	(mg)	235	5000	4.7
calcium	(mg)	7	900	0.8
phosphorus	(mg)	30	900	3.3
magnesium	(mg)		250	
iron	(mg)	0.6	14	4.3
water	(g)	93.9		

*RDA variable according to individual needs
**RDA recommended dietary allowance (average daily amount for a normal adult)

Sun dried tomato pasta sauce

(serves 4)

Ingredients		
1	package sun dried tomatoes	
45 mL	olive oil	3 TBSP
	2 medium onions or 8 shallots	
2-3	garlic cloves, minced	
60 mL	Marsala, port or sherry	1/4 cup
	parmesan cheese to taste	
	salt and pepper to taste	
	pasta of your choice	

Instructions

- Steep the sun dried tomatoes in warm water for about 30 minutes, then drain well.
- Heat oil, sauté minced onion and garlic for about 2 minutes.
- Add the tomatoes and wine, bring to the boil, and reduce heat. Stir until ingredients are well blended, adding a drop of water if necessary; season, serve with the pasta of your choice, and sprinkle on Parmesan cheese.

Portuguese plum tomato sauce

Ingredients		
1 kg	plum tomatoes	2 lbs
2	medium or large onions	
45 mL	olive oil	3 TBSP
2	garlic cloves, minced	
2	bay leaves	
3	cloves	
	squeeze of lemon juice	
5 mL	Worcestershire sauce	1 tsp
15 mL	fresh coriander, minced (optional)	1 TBSP
45 mL	port, sherry or wine	3 TBSP
5 mL	sugar	1 tsp
60 mL	water	1/4 cup
	salt and pepper to taste	
	parsley to garnish	

Instructions

- If desired, blanch the tomatoes to remove their skins. Slice them in rounds. Cut the onions in rings.
- In a deep skillet, heat the oil to medium and sauté the onions for 2 to 3 minutes without letting them brown.
- Add the tomatoes and garlic and mix thoroughly. After 2 minutes, add the rest of the ingredients. Simmer, covered, for about 10 to 15 minutes, stirring from time to time. Add a little water if necessary.
- Adjust seasoning, remove cloves and bay leaves, and garnish with minced parsley. Freeze for later use, if desired.
- This sauce is excellent on fish, especially grilled cod or tuna steaks, accompanied by spinach or Swiss chard and rice or potatoes. It is also good with pasta.

Jugoslavian mussels

(serves 4)

Ingredients		
2 kilos	fresh mussels	4 lbs
450 g	ripe plum tomatoes	1 lb
60 mL	olive oil	4 TBSP
6	garlic cloves, minced	
125 mL	parsley, finely chopped	1/2 cup
60 mL	flour	4 TBSP
30 mL	tomato paste	2 TBSP
30 mL	Vegeta*	2 TBSP
60 mL	grappa or brandy	4 TBSP
250 mL	white wine	1 cup

Instructions

- Clean the mussels well.
- Chop the plum tomatoes into small pieces. Heat the oil in a very large pot. Add the garlic, the parsley, the tomatoes and sauté for a minute or two.
- Add the mussels and stir for 2 or 3 minutes, until they are coated with the tomato mixture. Sprinkle on flour and stir until dissolved, then mix in the tomato paste and Vegeta*.
- Stir in the grappa, then the wine, then add enough boiling water to cover the mussels about half way up.
- Boil, then turn the heat down to medium. Cover and let cook for just a few minutes until the mussels have opened, stirring once or twice.
- Remove the pot from the heat, season to taste, then stir in a **large handful of breadcrumbs,** which will thicken the sauce nicely after a minute or so.
- Serve in bowls, accompanied by plenty of fresh, crusty bread to dip into the delicious sauce.

**Vegeta is a mixed dried spice that gives this dish a special flavour. It is widely used in Jugoslavia and Middle European countries, and is usually available at ethnic groceries. If unavailable, substitute 2 chicken or fish stock cubes, 30 mL (2 TBSP) dried mixed vegetable flakes and extra salt and pepper.*

Mini eggplant

Mini eggplant hors d'oeuvre

• One only has to look at mini eggplants to see why they have been named "eggplants", especially the white variety. These beautiful small purple or white vegetables, originally from tropical Asia, are members of the Solanaceæ family, which also includes tomatoes, peppers and potatoes. Eggplants were not known in Europe until the Moors introduced them in the 13th century. Today, most mini eggplants are grown in Holland.

• Mini eggplants have a creamy texture and a mild flavour that adapt well to a variety of recipes. They can be baked, grilled, fried, sautéed, stir-fried or stuffed with various mixtures, including spicy minced meat, rice, raisins and nuts, Mediterranean vegetables and shellfish.

• Choose smooth, glossy-skinned, unblemished eggplants for best results and use them within a few days.

Ingredients		
4	mini eggplants	
2-5 mL	salt	1/2 -1 tsp
15 mL	olive oil	1 TBSP
1	garlic clove, finely minced	
2	ripe tomatoes	
10 mL	white wine or herb vinegar	2 tsp
	salt and pepper to taste	
5 mL	chives, finely minced	1 tsp
2	fresh basil leaves, finely minced	

Instructions

• Cut the top off each eggplant and reserve them. If necessary, discard a thin slice from the base so they can stand upright.

• Carefully hollow out the flesh from each eggplant. Dice the flesh, sprinkle it with salt and let it stand for about 30 minutes. Pour off any liquid and dry the eggplant flesh with a paper towel.

• Heat the oil in a frying pan and sauté the eggplant flesh until it is tender, adding the garlic during this step.

• Skin the tomatoes (by blanching them), dice and add them to the frying pan.

• Add the white wine or herb vinegar to the mixture, season to taste, and continue cooking until the vegetables are ready.

• Cool, add the fresh herbs, fill the eggplant with the mixture and replace the caps.

• Serve as a cold hors d'oeuvre or a vegetable side dish.

Chestnut

Origin

• Edible chestnuts, often called Spanish chestnuts, are closely related to horse chestnuts, a wild variety that produces similar-looking nuts that are not eaten. Thousands — perhaps hundreds of thousands — of years ago, they grew in northern climates from Alaska to the Middle East, though today they are found in temperate climates and are susceptible to the cold. Several ancient Greek cities were famous for their magnificent chestnut trees. These trees were later introduced throughout Europe by the Romans, who adored roasted chestnuts.

• The cultivation of chestnuts was far more important before this century, when the chestnut was considered a staple food in Europe. Its composition is similar to other staple foods such as cereals (especially wheat), legumes and potatoes. For many centuries, chestnuts were ground into flour to make bread, particularly in the more remote parts of Western Europe. Today, chestnut flour is mainly used to make fancy cakes. As the potato and grains such as wheat and rice became more widespread in Europe, chestnuts became less important.

• Chestnut trees once grew abundantly in North America. Tragically, they were affected by blight at the end of the 19th century and were virtually made extinct. This majestic tree, once very important to the lumber trade, has now disappeared from many areas. Chinese chestnut trees have been planted in many parts of North America to replace the trees that died earlier.

• Chestnuts are considered good for the digestion but they should not be eaten by diabetics because of their sugar content.

• Chestnuts are imported from Portugal, Spain, Italy and France. Korea is another major supplier. Some chestnut trees also grow in Southern California.

Varieties and description

• The two most popular types of chestnut are the larger, more familiar and better-tasting European or Spanish chestnut and the small Chinese chestnut.

• The heart-shaped Spanish chestnut, flat on one side and rounded on the other, is a lovely shiny reddish-brown. It is rich, starchy and delicious. Nothing can be more tantalizing than the aroma of roasted chestnuts on a cold day.

GOOD SOURCE OF B VITAMINS
SOURCE OF IRON
LOW IN SODIUM

Availability

• Chestnuts are harvested from September to November. They are generally available through most of the winter.

Selection

• Choose chestnuts with smooth, firm, glossy shells that feel heavy for their size. Avoid those with dull, wrinkled shells.

Storage

• Keep chestnuts in a cool, dry place.

Preparation

• Chestnuts are too bitter to be eaten raw and nearly impossible to peel until they have been cooked. Cooking converts their starch into sugar, utterly transforming their taste. The shell, as well as the thin brown bitter skin, must be discarded.

• For quick preparation, try microwaving them; cut a slit into each chestnut, making sure you penetrate the nutmeat. Place them on a plate, uncovered, and microwave on high for 2 minutes. Turn them over and microwave for another minute, or until they are soft when squeezed.

• In Italy chestnuts are stewed in wine and in Portugal in port for a tasty side dish to accompany roasted fowl.

Suggestions

• Chestnuts are delicious roasted. Just make a light incision in the form of a cross on the rounded side and bake at 230 °C (450 °F) for 30 minutes or until they are tender. If you don't make proper incisions, they may explode. When they're cool enough to handle, peel and enjoy plain or in a recipe. Chestnuts can be roasted on an open fire, raised above the glowing embers, providing you use tongs to handle them.

• Try them boiled, whole or mashed with a little butter, salt and pepper. Add sautéed minced onions and a little garlic if desired; slit the shells and simmer for about 35 minutes until tender, then drain and cool before peeling and mashing them. A delicious side dish!

• Boiled chestnuts are excellent whole or sliced when mixed with cooked Brussels sprouts or spinach and a little butter and seasoning.

• Purée cooked chestnuts for use in desserts such as Nesselrode pudding, chestnut Bavarian cream and marrons Mont-Blanc as well as to flavour pastry cream for cake filling and fancy pastries.

• Puréed chestnuts are very good for thickening sauces and soups.

• For a gourmet dish to accompany roast chicken, duck or turkey, try chestnuts with wild rice. To serve 4, use 200 mL (3/4 cups) of wild rice, blanched for a few minutes in boiling water, then drained. Cook about 20 chestnuts for a few minutes, just until they can be peeled and crumbled into smaller pieces. Melt a little butter and gently sauté (until soft) 1 large onion, 2 celery stalks and 2 sprigs of fresh thyme, all finely minced. Add the rice, chestnuts, 250 mL (1 cup) soup stock and a good pinch of salt, pepper and nutmeg. Boil, then simmer about 30 minutes until the rice is ready, adding more liquid if necessary. (Wild rice does not absorb water the same way as ordinary rice and it is chewier.) Adjust seasoning, and serve on a bed of lightly cooked spinach for an attractive presentation.

Nutritional Value

Chestnut

(per 100 g = 3 1/2 oz)

			RDA**	%
calories*	(kcal)	191		
protein*	(g)	2.8		
carbohydrates*	(g)	41.5		
fat*	(g)	1.5		
fibre	(g)	1.1	30	3.7
vitamin A	(IU)		3300	
vitamin C	(mg)		60	
thiamin	(mg)	0.23	0.8	28.8
riboflavin	(mg)	0.22	1.0	22
niacin	(mg)	0.5	14.4	3.5
sodium	(mg)	2	2500	0.08
potassium	(mg)	410	5000	8.2
calcium	(mg)	29	900	3.2
phosphorus	(mg)	87	900	9.7
magnesium	(mg)		250	
iron	(mg)	1.7	14	12
water	(g)	52.5		

*RDA variable according to individual needs

**RDA recommended dietary allowance (average daily amount for a normal adult)

Chestnut soup

(serves 4)

Ingredients

450 g	chestnuts	1 lb
500 mL	soup stock	2 cups
125-250 mL	milk	1/2-1 cup
1 pinch	each ground cloves, nutmeg and sugar	
30 mL	butter or margarine	2 TBSP
1	medium onion, minced	
3 stalks	of celery, minced	
2 sprigs	fresh thyme	
1	bay leaf	
	salt and pepper to taste	
	light cream (optional)	
	parsley for garnish, finely minced	

Instructions

• Slit the chestnuts then boil, roast or microwave them for a few minutes until they can be shelled. (They are easier to peel while still warm.)

• Place the peeled chestnuts in a pot with vegetable or chicken stock, adding enough milk so they are well covered. Add pinches of ground cloves, nutmeg and sugar.

• Gently simmer for about 20 minutes or until tender.

• Melt the butter in a frying pan and gently sauté the onions, celery, thyme and bay leaf over a low heat for 10 to 15 minutes, covered, until very soft but not browned. Discard the thyme and bay leaf.

• Add the vegetables to the chestnuts and purée everything in a blender.

• Heat the soup gently, seasoning with salt and pepper to taste.

• Add a little milk or light cream if the soup is too thick.

• Garnish each bowl with a sprinkling of finely minced parsley and serve.

Chestnut stuffing for turkey

Ingredients		
500 mL	chestnuts	2 cups
45 mL	butter or olive oil	3 TBSP
250 mL	onions, minced	1 cup
500 mL	celery, minced	2 cups
5 mL	celery salt	1 tsp
	handful of chopped parsley	
2	chicken stock cubes	
750 mL	whole wheat breadcrumbs*	3 cups
	juice of 1 orange	
125 mL	raisins	1/2 cup
	water to moisten	
	salt and pepper to taste	

Instructions

- Make a slit in each chestnut, then boil them for at least half an hour, or until tender. Peel chestnuts and break into pieces.
- In a large frying pan, heat butter or oil. Sauté onion and celery with celery salt. Cook until soft.
- Add parsley and crumbled stock cubes, stir well. Add breadcrumbs and mix well.
- Add orange juice, chestnuts, raisins and enough water to moisten the mixture, until it sticks together.
- Season to taste. Use to stuff turkey or chicken.
- This stuffing can be prepared ahead of time and frozen.

**Substitute 250-375 mL (1-1½ cups) cooked quinoa (a small South American grain) or 250 mL (1 cup) cooked wild rice for an equivalent amount of breadcrumbs, if desired.*

Durian and jackfruit

↑ Durian

Origin

• Although these two tropical prehistoric-looking fruits are not related, they seem alike to the uninitiated. They are surely among the largest and oddest fruits on earth.

• The durian tree is related to the baobab and the kapok trees. Native to the Malaysian area, this outlandish, extremely smelly fruit, also known as civet fruit, is exceptionally popular throughout Southeast Asia. Scientists are working on producing a durian that doesn't smell, which would hasten its acceptance in the West. According to a Malaysian saying, the fruit is paradise but its smell is hell. Many Asians believe it to be an aphrodisiac; a local aphorism alleges: "When the durian falls, the skirt rises." The word durian literally means thorn fruit in Malaysian. Today, Thailand and Malaysia are among the main producer countries, along with Indonesia and the Philippines.

• The jackfruit, which comes from the botanical family that includes the fig and the mulberry, is closely related to breadfruit and marang. Jackfruit was known to the ancient Romans as one of the curiosities of the East.

JACKFRUIT IS A FAIRLY GOOD SOURCE OF VITAMIN C AND MAGNESIUM
VERY LOW IN SODIUM AND FIBRE

• Indigenous to areas of the South Pacific, Southeast Asia and parts of India, jackfruit is grown extensively throughout the East and in some regions of Africa, including Kenya. The Portuguese introduced jackfruit to Brazil in the 17th century, and from there it spread to various islands in the Caribbean. Jackfruit has the advantage of having a very mild odour compared with durian.

Varieties and description

• The durian, a fruit that strongly resembles a medieval weapon, is large, globular and covered with sharp spikes. It usually weighs between 1.5 and 6 kilos (3 and 13 pounds), though it can be much larger. Durians are typically 30 centimetres (1 foot) or more in diametre. The thorny spines that cover its thick skin, which can be yellowish green to light brown, make the fruit difficult to handle without gloves. Workers harvesting durian are advised to wear helmets, as a fruit falling from a distance can be lethal.

• Several faint grooves in its skin indicate the lines on which the fruit should be opened; there usually are five or six separate sections, divided by inedible white membranes, each containing two or three seeds embedded in sweet, creamy-textured yellow flesh, which lifts out in irregularly shaped sections. Opinions on the taste of a durian vary wildly, from a combination of strawberries, bananas, custard and cream to shallots or garlic flavoured with vanilla! Its notorious smell, which worsens as the fruit ripens, has led to durian being banned from hotels, airlines and most forms of public transportation. Although East Asians are wild about durian, it is definitely an acquired taste for a Westerner.

• Jackfruit are similar in appearance to durians, though usually much larger and more elongated. The texture and odour of the fruit is different and more agreeable to a western palate. Although a typical fruit weighs from 7 to 15 kilos (15 to 33 pounds), they are known to grow up to about 40 kilos (nearly 90 pounds). The most popular varieties of jackfruit have a hard skin or rind. Pale green or yellowish-green, the skin turns brownish as it ripens and is covered with hard, protruding scales. Their sweet flesh can be yellow, orange, creamy white or even pink, depending on the variety. Jackfruit contains numerous seeds that should never be eaten raw.

Availability

- Durians are usually available from April to August.
- Jackfruit are available throughout the year, as they are grown in so many countries. Thailand, a major producer, has its harvest from January to May.

Selection

- Durians are exported by air in an unripe state, and as long as the skin has not been punctured, they should be good.
- Choosing a good jackfruit is not simple. But because of their large size, they are often sold in pieces, which makes it easier to tell if the fruit is good.
- Though the odour of both fruits is pronounced, they should never smell fermented.

Storage

- A durian should ripen in a few days, if left at room temperature. Any signs of cracking in the skin indicates ripeness. They are best eaten as soon as they are ripe.
- Jackfruit that has been imported is usually kept about a week or more at room temperature to ripen. Once the fruit has been opened and removed from its rind, it can be kept for a few days in the refrigerator.

↓ Jackfruit

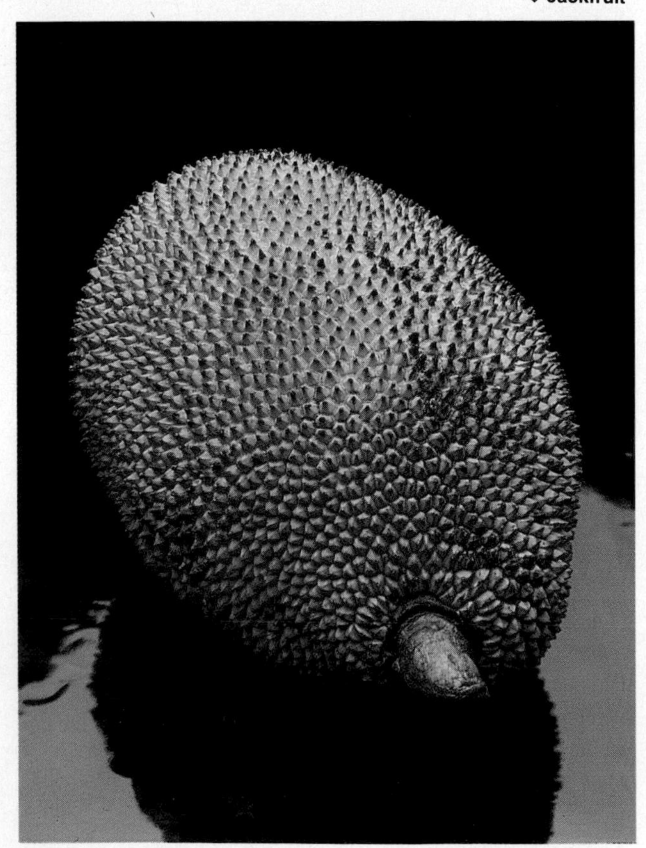

Preparation

- As a durian ripens, its thick, tough skin begins to crack open along its "seams". A very sharp knife should be used to open the fruit fully. Once opened, any leftover durian should be removed from its skin, wrapped well to prevent the odour from permeating other foods and refrigerated. Its seeds, like those of the jackfruit, can be treated like chestnuts and roasted or boiled.
- When opening a jackfruit, it is best to oil your knife and possibly your fingers to prevent its sticky white sap from adhering. Cut it in quarters lengthwise, discard the core, then invert each quarter to separate the small "fruitlets", each of which contains a large seed. Pull out each fruit section, slit open and remove the seed.

Suggestions

- Durians are usually eaten plain, as fresh as possible. From Thailand to Indonesia, durians have traditionally been eaten with "sticky rice" and sweet coconut milk. In recent years, they have been used to make moon cakes, tarts, jam, waffles and ice cream.
- Jackfruit are usually eaten in their natural state or lightly cooked in syrup. They are sometimes added to a curry or even fried. In Java, they are boiled in coconut milk. Jackfruit is good in fruit punch or served with ice cream.
- Neither fruit should be prepared or consumed with alcohol, as the combination may cause a chemical reaction that has unpleasant effects on some people.

Nutritional Value

Jackfruit

(per 100 g = 3 1/2 oz)

			RDA**	%
calories*	(kcal)	94		
protein*	(g)	1.5		
carbohydrates*	(g)	24		
fat*	(g)	0.3		
fibre	(g)	1	30	3.3
vitamin A	(IU)	297	3300	9
vitamin C	(mg)	7	60	11.7
thiamin	(mg)	0.03	0.8	3.8
riboflavin	(mg)		1.0	
niacin	(mg)	0.4	14.4	2.8
sodium	(mg)	3	2500	0.12
potassium	(mg)	303	5000	6.1
calcium	(mg)	34	900	3.8
phosphorus	(mg)	36	900	4
magnesium	(mg)	37	250	14.8
iron	(mg)	0.6	14	4.3
water	(g)	72.3		

*RDA variable according to individual needs

**RDA recommended dietary allowance (average daily amount for a normal adult)

Index of Fruits and Vegetables

chervil	cerfeuil	perifolio	*Anthriscus cerefolium*
chive	ciboulette	cebollino	*Allium schoenoprasum*
coriander/cilantro	coriandre	cilantro	*Coriandrum sativum*
dill	aneth	eneldo	*Anethum graveolens*
garlic	ail	ajo	*Allium sativum*
marjoram	marjolaine	mejorana	*Origanum majorana*
mint	menthe	menta	*Mentha* spp.
oregano	origan	oregano	*Origanum* spp.
parsley	persil	perejil	*Petroselinum* spp.
rosemary	romarin	romero	*Rosmarinus officinalis*
sage	sauge	salvia	*Salvia officinalis*
savory	sariette	ajedrea	*Satureja hortensis*
tarragon	estragon	tarragona	*Artemisia dracunculus sativa*
thyme	thym	tomillo	*Thymus vulgaris*
Honeydew, *76*	Melon miel	Melòn de miel	*Cucumis melo*
Honey tangerine/Murcott, *93, 207*	Tangerine au miel/Murcott	Mandarina de miel	*Citrus reticulata*
Horned melon, *160*	Melon à cornes	Melón cornudo	*Cucumis metuliferus*
Horseradish, *180*	Raifort	Rábano picante	*Armoracia rusticana*
Jackfruit, *251*	Jaque	Jaca	*Artocarpus heterophyllus*
Jerusalem artichoke, *177*	Topinambour	Alcachofa jerusalen	*Helianthus tuberosus*
Jicama, *42*	Jicama	Jícama	*Pachyrrhizus erosus*
Kale, *174*	Borécole	Berza col/col rizada	*Brassica oleracea, var. acephala f. sabellica*
Kiwi/Chinese gooseberry, *7*	Kiwi/groseille de Chine	Kiwi	*Actinidia chinensis*
Kohlrabi, *170*	Chou-rave	Colirrámbano	*Brassica oleracea, var .gongylodes*
Kumquat, *210*	Kumquat	Kumquat	*Fortunella polyendra*
Lemon, *202*	Citron	Limón	*Citrus limonum*
Lemon grass, *117*	Jonc odorant	Limoncillo	*Cymbopogon citratus*
Lettuce, hydroponic, *36*	Laitue hydroponique	Lechuga hydroponica	*Lactuca sativa*
Lime, *202*	Lime/citron vert	Lima	*Citrus latifolia/C. aurantifolia*
Longan, *100*	Longane	Longana	*Euphoria longana*
Long melon/opo, *124*	Melon long/opo	Cojombro/Guiro amargo	*Lagenaria siceraria*
Loquat, *152*	Nèfle/bibasse/loquat	Nispero del japon	*Eriobotrya japonica*
Lotus root, *123*	Racine de lotus	Raíz de loto	*Nelumbo nucifera*
Lychee, *100*	Litchi	Litchi	*Litchi chinensis*
Mâche/lamb's lettuce, *50*	Mâche/valérianelle	Macha	*Valerianella locusta*
Malanga, *130*	Malanga	Yautia/malanga	*Xanthosoma sagittifolium*
Mandarin, *93, 206*	Mandarine	Mandarina	*Citrus reticulata*
Mango, *90, 134*	Mangue	Mango	*Mangifera indica*
Mangosteen, *5*	Mangoustan	Mangostan	*Garcinia mangostana*
Melon, *73, 76, 78, 79, 82*	Melon	Melón	*Cucumis melo*
Mungbean sprouts, *56*	Haricot mungo germé	Tallos de mungo	*Vigna radiata*
Mushrooms, *114, 187*	Champignons	Hongos	
chanterelle, *187*	chanterelle	hongo chantarél	*Cantharellus cibarius*
enoki, *114*	enoki	hongo enoki	*Flamulina velutipes*
morel, *187*	morille	hongo morél	*Morchella* spp.
oyster mushroom, *114*	pleurote	hongo pleurote	*Pleurotus ostreantus*
porcini/bolete, *187*	cèpe	hongo porcini	*Boletus edulis*
shiitake, *114*	shiitake	hongo shiitake	*Lentinus edodes*
wood ear, *114*	oreille de judas	hongo chino	*Auricula* spp.
Mustard cress/Garden cress, *56*	Germes de moutarde	Tallos de mostaza	*Lepidium sativum*
Nappa, *111*	Nappa	Lechuga orejana	*Brassica pekinensis*
Okra, *190*	Gombo/okra	Quimbombó	*Abelmoschus esculentus/ Hibiscus esculentus*
Papaya, *12, 134*	Papaye	Papaya	*Carica papaya*

Passion fruit, *137*	Fruit de la passion	Maracuya/granadilla	*Passiflora edulis/P. edulis flavicarpa*
Pear, *216*	Poire	Pera	*Pyrus communis*
Pearl onion, *193*	Oignon perlé	Cebollín	*Allium ampeloprasum, var. holmense*
Pepino/melon pear, *86*	Pepino	Cachun	*Solanum muricatum*
Pepper (sweet), *52*	Poivron	Pimienta	*Capsicum annuum*
Persimmon, *143*	Kaki/plaquemine	Caqui	*Diospyros kaki*
Physalis/Cape gooseberry, *146*	Physalis/groseille du Cap	Uvilla/Capuli	*Physallis peruviana*
Pineapple, *2*	Ananas	Piña	*Ananas comosus*
Pitahaya, *140*	Pitahaya	Pitahaya	
Plantain, *230*	Plantain	Plátano	*Musa* spp.
Pomegranate, *156*	Grenade/pomme-grenade	Granada	*Punica granatum*
Pomelo, *212*	Pomélo	Pomelo	*Citrus grandis/C. maxima*
Prickly pear/cactus fruit, *140*	Poire cactus/figue de Barbarie	Tuna	*Opuntia tuna/O. ficus-indica*
Prickly pear pads, *140*	Nopale	Nopales	*Nopalea spp.*
Purslane, *64*	Pourpier	Verdolaga	*Portulacca oleracea*
Quince, *149*	Coing	Membrillo	*Cydonia oblonga*
Radicchio, *46*	Radicchio	Radicchio	*Chicorium intybus, var. foliosum*
Radish, oriental/daikon, *113*	Radis oriental/daïkon	Rábano oriental	*Raphanus sativus longipinnatus*
Radish sprouts, *56*	Germes de radis	Tallos de rábano	*Raphanus sativus*
Rambutan, *10*	Ramboutan	Rambutan	*Nephelium lappaceam*
Rapini, *28*	Rapini		*Brassica rapa*
Rhubarb chard, *182*	Carde-rhubarbe	Açelga suiza roja	*Beta vulgaris, var. cicla*
Salak, *227*	Salak	Salak	*Zalacca edulis*
Salsify, black, *235*	Salsifis noir	Salsifi negro	*Scorzonera hispanica*
Salsify, white, *235*	Salsifis blanc	Salsifi blanco	*Tragopogon porrifolius*
Savoy salad, *34*	Laitue savoy	Lechuga savoy	*Brassica oleracea acephala*
Scallion/green onion	Oignon vert	Cebolleta	*Allium cepa*
Seville orange, *204*	Bigarade	Naranja agria	*Citrus aurantium bigardia*
Shallot, dry, *193*	Échalote sèche	Chalote ascalonia	*Allium ascalonicum*
Snow pea, *108*	Pois mange-tout	Vainita china	*Pisum sativum macrocarpon*
Sorrel, *62*	Oseille	Acedera	*Rumex acetosa, var. hortensis*
Soursop, *166*	Corossol	Guanabana	*Annona muricata*
Soybean sprouts, *56*	Germes de soja/soya	Tallos de soya	*Glycine max*
Sprouts, *56*	Germes		
Squash, *196*	Courge	Calabaza	*Cucurbita pepo, C. maxima C. mixta C. moschata*
acorn	poivrée/courgeron		
banana	banana		
buttercup	buttercup		
butternut	musquée		
calabaza	calebasse		
cocozelle	coucourzelle		
crookneck	cou tors		
delicata	delicata		
dumpling	dumpling		
golden nugget	golden nugget		
Hubbard	de Hubbard		
kabocha	kabocha		
patty pan	patisson		
pumpkin	citrouille		
spaghetti	spaghetti		
straightneck	cou droit		
turban	giraumon/turban		
zucchini	courgette/zucchini		
Sweet potato, *239*	Patate douce	Batata/patata dulce	*Ipomoea batatas*

Swiss chard, *182*	Bette à carde	Açelga suiza	*Beta vulgaris, var. cicla*
Tamarillo, *154*	Tamarillo	Tomate de la Paz	*Cyphomandra betaceae*
Taro/dasheen/eddo, *130*	Taro/Dachine/Eddo	Taro	*Colocasia esculenta*
Tomatillo, *146*	Tomatillo	Tomatillo	*Physallis ixocarpa*
Tomato, *48, 241*	Tomate	Tomate	*Lycopersicon esculentum*
Tomato,plum, *241*	Tomate prune	Tomate barceló	*Lycopersicon esculentum*
Uglifruit, *212*	Ugli	Ugli	*Citrus reticulata* spp.
Water chestnut, *136*	Châtaigne d'eau	Cabezas de negrito	*Eleocharis dulcis*
Watercress, *58*	Cresson	Berro	*Nasturtium officinale/N. aquaticum*
Watermelon, *88*	Melon d'eau	Sandia	*Citrullus lanatus*
Winter melon, *124*	Melon d'hiver	Calabaza china	*Benincasa hispida*

AVAILABILITY CHART

	J	F	M	A	M	J	J	A	S	O	N	D
Apple, Braeburn	Unavailable	Unavailable	Unavailable	Available	Available	Available	Available	Available	Unavailable	Unavailable	Unavailable	Unavailable
Apple, Gala and Royal Gala	Unavailable	Unavailable	Unavailable	Available	Available	Available	Unavailable	Unavailable	Unavailable	Unavailable	Unavailable	Unavailable
Apple, Gala (U.S. & Can.)	Unavailable	Unavailable	Unavailable	Unavailable	Unavailable	Unavailable	Unavailable	Unavailable	Available	Available	Available	Available
Apple pear	Plentiful	Plentiful	Available	Available	Available	Available	Available	Available	Plentiful	Plentiful	Plentiful	Plentiful
Artichoke	Available	Available	Available	Plentiful	Plentiful	Available	Available	Available	Available	Plentiful	Available	Available
Arugula	Available	Available	Available	Available	Available	Available	Available	Available	Available	Available	Available	Available
Asparagus	Available	Available	Plentiful	Plentiful	Plentiful	Plentiful	Available	Available	Available	Available	Available	Available
Avocado	Available	Available	Available	Available	Plentiful	Plentiful	Plentiful	Plentiful	Available	Available	Available	Available
Banana, unusual varieties	Available	Available	Available	Available	Available	Available	Available	Available	Available	Available	Available	Available
Cantaloupe	Available	Available	Available	Available	Available	Plentiful	Plentiful	Plentiful	Available	Available	Available	Available
Carambola/starfruit	Available	Available	Available	Available	Available	Available	Available	Available	Available	Available	Available	Available
Cardoon	Available	Available	Available	Available	Available	Unavailable	Unavailable	Unavailable	Available	Available	Available	Available
Celeriac	Plentiful	Plentiful	Plentiful	Plentiful	Available	Available	Available	Available	Available	Plentiful	Plentiful	Plentiful
Champagne grapes	Unavailable	Unavailable	Unavailable	Unavailable	Unavailable	Unavailable	Available	Available	Available	Available	Unavailable	Unavailable
Chanterelle	Unavailable	Unavailable	Unavailable	Unavailable	Unavailable	Available	Available	Available	Available	Available	Available	Available
Chayote	Plentiful	Plentiful	Plentiful	Available	Available	Available	Available	Available	Available	Available	Available	Plentiful
Cherimoya	Available	Available	Available	Available	Available	Available	Available	Available	Available	Available	Available	Available
Cherry, Montmorency	Unavailable	Unavailable	Unavailable	Unavailable	Unavailable	Unavailable	Available	Available	Unavailable	Unavailable	Unavailable	Unavailable
Cherry, Rainier	Unavailable	Unavailable	Unavailable	Unavailable	Available	Available	Available	Unavailable	Unavailable	Unavailable	Unavailable	Unavailable
Chestnut	Plentiful	Plentiful	Unavailable	Unavailable	Unavailable	Unavailable	Unavailable	Unavailable	Plentiful	Plentiful	Plentiful	Plentiful
Chili peppers	Available	Available	Available	Available	Available	Plentiful	Plentiful	Plentiful	Available	Available	Available	Available
Chinese greens	Available	Available	Available	Available	Available	Available	Available	Available	Available	Available	Available	Available
Coconut	Plentiful	Available	Available	Available	Available	Available	Available	Available	Available	Plentiful	Plentiful	Plentiful
Collards	Available	Available	Available	Available	Available	Unavailable	Unavailable	Unavailable	Unavailable	Available	Available	Available
Cranberry	Available	Unavailable	Unavailable	Unavailable	Unavailable	Unavailable	Unavailable	Unavailable	Available	Available	Available	Available
Dandelion	Available	Available	Available	Available	Available	Available	Available	Available	Available	Available	Available	Available
Durian	Unavailable	Unavailable	Unavailable	Available	Available	Available	Available	Available	Unavailable	Unavailable	Unavailable	Unavailable
Endive	Available	Available	Available	Available	Available	Unavailable	Unavailable	Unavailable	Available	Available	Available	Available
Fig	Available	Available	Available	Available	Available	Plentiful	Plentiful	Plentiful	Plentiful	Plentiful	Available	Available
Feijoa	Unavailable	Unavailable	Available	Available	Available	Available	Unavailable	Unavailable	Available	Available	Available	Unavailable
Fennel	Available	Available	Available	Available	Available	Available	Available	Available	Available	Available	Available	Available
Fiddleheads	Unavailable	Unavailable	Unavailable	Available	Available	Available	Unavailable	Unavailable	Unavailable	Unavailable	Unavailable	Unavailable
Fresh herbs	Available	Available	Available	Available	Available	Available	Available	Available	Available	Available	Available	Available
Ginger	Available	Available	Available	Available	Available	Available	Available	Available	Available	Available	Available	Available

Availability of fresh produce is subject to weather, supply and demand.

Plentiful — Available — Unavailable

AVAILABILITY CHART

	J	M	M	A		J	J	A	S	O	N	D
Guava	A	A	A	A	A	A	A	A	A	A	A	A
Honeydew, green	A	A	A	A	A	P	P	P	P	P/A	A	A
Honeydew, orange	A	A	A	A	A	U	A	A	A	U	U	A
Horned melon	U	U/A	A	A	A	U	U	U	U	U	U	U
Horseradish	A	A	P	P	P	A	A	A	A/P	P	P	P/A
Jackfruit	A	A	A	A	A	A	A	A	A	A	A	A
Jerusalem artichoke	P	P	P	A	A	A	A	A	P	P	P	P
Jicama	P	P	P	P	A	A	A	A	A	A	A	P
Kale	P	P	P	P	A	A	A	A	A	A	A	P
Kiwi	A	A	A	A	A	A	A	A	A	A	A	A
Kohlrabi	A	A	A	A	A	P	P	P	P	P	A	A
Kumquat	P	P	P	P	A	A	A	A	A	A	P	P
Lettuce, hydroponic	A	A	A	A	A	A	A	A	A	A	A	A
Lime	A	A	A	A	A	P	P	P	A	A	A	A
Longan	U	U	U	U	U	U	U	A	A	U	U	U
Longbean	A	A	A	A	A	A	A	P	P	P	A	A
Loquat	U	U	A	A	A	A	U	A	A	A	U	U
Lychee	A	A	A	A	U	A	P	P	A	A	A	A
Mâche	P	P	P	P/A	A	A	A	A	A	A	A	P
Mandarin hybrids	A	A	A	A	A/U	U	A	A	A	A	A	A
Mango	A	A	A	A	P	P	P	P	A	A	A	A
Mangosteen	A	A	A	U	A	A	A	A	A	U	U	A
Morel	U	U	U	A	A	A	A	U	A	U	U	U
Nappa	A	A	A	A	A	A	A	A	A	P	P	P
Okra	A	A	A	A	A	P	P	P	P	P	A	A
Orange, blood	A	A	A	U	U	U	U	U	U	U	A	A
Orange, Seville	A	A	A	U	U	U	U	U	U	U	U	A
Oriental eggplant	A	A	A	A	A	A	A	A	A	A	A	A
Oriental mushrooms	A	A	A/P	P	P	P/A	A	A	P	P	P	A
Oriental radish	A	A	A	A	A	A	A	A	A	A	A	A
Oriental squash	A	A	A	A	A	A	P	P	P	P	A	A
Papaya	A	A	A	A	A	A	A	A	A	A	A	A
Passion fruit	A	A	P	P	P	P	P	P	P	P	A	A
Pear	A	A	A	A	A	A	A	A	A	A	A	A
Pearl onions	A	A	A	A	A	A	A	A	A	A	A	A

Availability of fresh produce is subject to weather, supply and demand.

P = Plentiful A = Available U = Unavailable

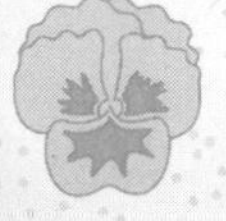

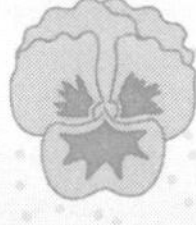
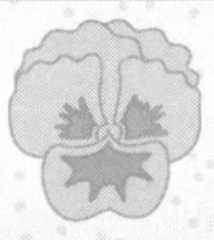
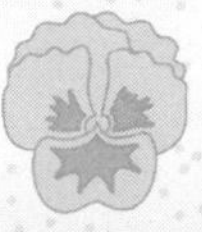
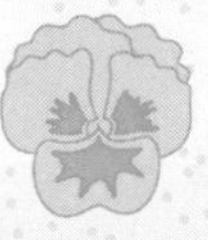
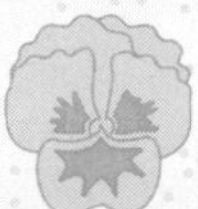